William Emerson Damon

Ocean Wonders

A Companion for the Seaside

William Emerson Damon

Ocean Wonders
A Companion for the Seaside

ISBN/EAN: 9783337042684

Printed in Europe, USA, Canada, Australia, Japan

Cover: Foto ©berggeist007 / pixelio.de

More available books at **www.hansebooks.com**

OCEAN WONDERS:

A

COMPANION FOR THE SEASIDE.

FREELY ILLUSTRATED FROM LIVING OBJECTS.

BY

WILLIAM E. DAMON.

"We know not a millionth part of the wonders of this beautiful world."

NEW YORK:
D. APPLETON AND COMPANY,
549 AND 551 BROADWAY.
1879.

TO

THE LOVERS AND ADMIRERS

OF

THE OCEAN AND ITS WONDERS

THIS VOLUME IS DEDICATED BY

THE AUTHOR.

PREFACE.

THIS volume has been specially prepared with a view to supply a long-felt need of precise and reliable information in regard to the living objects of our own sea-coast, and, incidentally, of other marine animals, either suitable for the aquarium or of sufficient intrinsic interest to deserve notice in any general work on marine zoölogy. My purpose is not only to describe the organic features and modes of life of these creatures, in their native ocean and river homes, but also to give careful and practical instruction as to *where* and *how* many of them may be procured and preserved in parlor and public aquaria.

Thousands of persons visit the sea-shore, perhaps spending weeks annually in its vicinity, without any idea of availing themselves of these opportunities to become practically acquainted with the curious and interesting forms of life which abound in such localities. Nature throws her choicest treasures at their feet, but they walk over them disregardful and insensible; while it is true that some even of the commonest productions of the sea—productions which are unnoticed from their very abundance—would well repay careful study and patient investigation. The invisible is often more curious than that which meets the eye, and a microscope of even moderate power will reveal in sea-water the marvelous resources of Nature in furnishing the minutest creations with organs as perfectly adapted to their mode of life as are those of the gigantic whale or the vertebrate land animals.

The realm of marine zoölogy is very wide-reaching, and includes such a vast variety of large and small, simple and complicated organisms as to be practically limitless. No one human life is long enough to compass a thorough acquaintance with all the "wonders of the ocean," concealed as many of them are in the depths of that immense and prolific reservoir of animated existence. But, from the unnumbered total, it is my purpose in this work to draw forth examples of each important class, and I shall endeavor to give mainly such information as will tend to excite increased interest in the subject, and will enable those whose curiosity is thus aroused to effect a personal acquaintance with such specimens as are obtainable on our own shores or the neighboring West India Islands.

Certain it is that the more intimate one becomes with this department of science the broader it grows, until the student is not only lost in wonder at the inexhaustible varieties of life displayed, but in endless reflections upon the uses of such profusion and such complications of means to meet the wants of apparently insignificant creatures.

Often do I think of a remark made by my friend the late Dr. Edmund R. Peaslee, while inspecting my collection of minute marine animals, that "the thorough investigation of any branch of natural science always ends in a mystery."

The remark was true; but, before reaching the mystery, how much pleasure and instruction are derived from the mere investigation! The study of the wondrous formations of sea-polyps, of the delicate alcyonarian zoöphytes and corals, of the sea-urchins, the lily-stars, and the strange fish of delicate form and tint, opens to the student who approaches the subject for the first time an entirely new world of unexpected pleasure and unimagined forms of existence. The sea, to him, never more can be simply a vast stretch of navigable water, useful as a highway for commerce, or as a means of summer recreation.

Of books upon aquaria—particularly of English reprints —there has been a large number given to the American public. Many of them are valuable and interesting in their way, but they are generally deficient in that sort of information which the amateur *most needs*, namely, direct and comprehensible instruction "how to keep an aquarium." Some of the attempts to do this are altogether *misleading* and *impracticable*, while others are not adapted to the exigencies of our climate. Some of the most learned and scientific writers, both American and foreign, fail lamentably on this important point. It is not so easy as it appears at the first glance to assure success in establishing a private aquarium. Whatever value this volume possesses is due to the fact that I give no second-hand directions, but the results and deductions of my own dearly-bought personal experience, attained at a considerable outlay not only of time and trouble but also of money, in obtaining many rare and scarce specimens of marine life, and in experiments to ascertain the kind of animals which would survive captivity. In the latter, I hope my directions or hints will materially diminish the amount of expenditure for such amateurs as may peruse this book. But even the process of learning by defeats and failures has its reward, and, when success is won, the aquarium becomes a permanent pleasure—a "joy forever!"

I must, however, warn the reader, particularly the novice in natural science, that he must not expect in a volume of this size anything like a complete description, or a mention even, of every specimen which may be found upon our shores. Neither must he imagine that the excellent illustrations which accompany most of the varieties introduced to his notice can adequately exhibit the delicate beauty and exquisite, changeful tints of the living animal. In many cases these are as evanescent as they are beautiful, and neither pen nor pencil can convey a full and satisfactory representation. We can give the form and approximate to the shades of color, but the beauty inherent in vitality is untransferable to the printed page. So perfectly inexhaustible are these minute objects, that one may spend *hours* or *years* of never-ending pleasure and interest upon them.

The preëminent value of studies in natural science, pursued either as a profession or as the recreation of a busy life, will not be questioned at this late day. A practical knowledge of any branch of natural science, and an interest in it, are conceded to be the best possible tonic for producing physical and mental health and vigor. Of the amount of absolute pleasure to be derived from such pursuits, every enthusiastic amateur is a living illustration. The experimental naturalist is almost invariably a happy man.

In conclusion, I desire to record my obligation to those friends who have assisted me by sympathy or otherwise in this work. And, first of all, to my dear and honored sister, whose suggestive spirit and practical example awakened in my mind a love for this charming science, I here express my affectionate gratitude for the introduction she gave me to a knowledge of aquarial life, feeling that to her, and to the intelligent assistance and sympathetic interest of my wife in my favorite studies, I am indebted for some of my happiest hours.

My valued friend Robert A. West—would that he still lived to receive my acknowledgments!—was the most de-voted lover of aquarial science that I have ever met. Being connected editorially with the " Commercial Advertiser " of this city, he exerted perhaps more influence than any other individual in advancing the subject in the public appreciation. On his practical wisdom and counsel I could always rely.

H. Dorner, Ph. D., late of the Hamburg Zoölogical Garden and Aquarium; Prof. H. D. Butler, and Messrs. Charles Reiche & Brother, of this city, have also courte-ously and kindly afforded me assistance in various ways.

It would be a long list should I undertake to name all the kind friends who, in the course of years, have aided me in my aquarial pursuits; but, though space does not permit me to mention them, not one is forgotten.

In connection with the publication of this volume, I cannot in justice omit the name of J. Carson Brevoort, Esq., the eminent scholar and scientist, late Superintendent of the Astor Library, to whom I am indebted for valuable suggestions and the loan of authorities; nor that of Mrs. E. Vale Blake, for special assistance in the revision of these pages for the press.

W. E. D.

NEW YORK, *January*, 1879.

CONTENTS.

ILLUSTRATIONS.

THE OCEAN.

CHAPTER I.

INTRODUCTORY.

THE ocean! the vast, glorious, boundless blue! How the vision of sunny hours, inspiring breezes, the invigorating scent of the salt air, and the sparkling of bright sea-foam, rises at thought of the great deep—that restless, deceptive, yet ever-enchanting siren, which lures us in every tone of the gamut to trust ourselves on its sparkling bosom! When sunny and serene, how seductive, how harmless it looks; when lashed into bright foam by contending wind and tide, what various tints and shades of beauty creep up the crystal walls of the many-headed billows! Watched from the shore, it is charming in its serenity, grand and glorious even in its wildest fury. But, sailing on its surface, how the wonders of the ocean deepen and expand!

In some parts of this vast water-system we should find it much salter than in others, and should perceive that other differences exist, not only in the proportions of its chemical ingredients, but in its specific gravity, its color, its purity, its thermal gradations, and innumerable other variations.

In one part of the ocean a traveler might be endangered by icebergs, in another becalmed, as in the "doldrums," or

entangled in the giant algæ of the Sargasso Sea. And he
would learn that he could scarcely sail in a perfectly direct
course from any given point to another, but that he must
more or less obey the tides, currents, and prevailing winds.
He might even vary his diversion by sailing up and down
rivers in the sea, should he chance to navigate the long axis
of the Gulf Stream or the great current of Japan.

Then, again, in some parts of the ocean, if he looked over
the bulwarks at night, he would see diffused or trailing lights,
as if the milky-way had descended from its sphere and was
floating on the sea, or as if the sun had left some broken rays
of its departing glory on the waters, ere it descended below
the horizon. On the surface, again, he might encounter the
larger cetacea, whales, and schools of porpoises, disporting
themselves; or an immense growth of *fuci*, which he would
naturally report as a "sea-serpent;" or see dolphins chasing
flying-fish, while robber birds disputed with them for the
prey. These and many other interesting sights would help
to charm away the hours, while the reflection that millions
of the human race depend for their existence mainly upon
the products of the ocean would add immense interest to all
that related to it. Nature seems to have bestowed upon the
most valuable edible inhabitants of the ocean a capacity for
increase simply astounding; but it is not food alone which
the sea furnishes to man. Oil for lights and for mechanical
uses, chemical substances for dyes and drugs, valuable furs
for clothing, and many other useful and beautiful objects,
would render the ocean an almost exhaustless subject for
study, did there remain no others than those which we have
named.

But these various conditions of the ocean and the objects
to be seen upon its surface are not the "wonders" of which
we mean to speak, but rather of that vast reservoir of life
hidden beneath these beautiful waves, of those curious and
charming marine animals of which one can scarcely realize

the attraction, unless, like the writer, he has watched their modes of life with attention, and observed from day to day the wonderful provision which Nature has made for the growth and sustenance of even the minutest organisms. The immense variety of life contained in the ocean would never be suspected from superficial observations. The very scintillations of the surface, which we admire at night, are but the myriad lamps of tiny creatures, wonderfully fashioned, and serving purposes which eventually, by circuitous paths, redound to the advantage of the human race. As purifiers of the ocean by their constant movements, and as food for larger kinds of marine animals, they prove as serviceable to man as they are eminently worthy of his study.

The only way by which amateur naturalists can familiarize themselves with living specimens of marine animals is by collecting them in aquaria. These artificial reservoirs are looked upon by many as a modern fashionable invention; but history informs us that the art of preserving and breeding rare marine and fresh-water animals was well understood by the ancients. The Romans, in those luxurious days which preceded the decline of the empire, were famous for the lavish expenditure bestowed upon their artificial fish-ponds; and some very ingenious and curious means were adopted by them in the preservation of aquatic animals. True, it was not with a view to scientific observation, or even as amateur naturalists, but rather with the epicurean intent to procure the choicest fish for their tables. But in some instances, doubtless, it was for the beauty of the creatures themselves, as in the case of the gold-fish and the red mullet, the latter of which displays in death those marvelous iridescent colors which induced some of the patricians to introduce streams of water under their dining-tables, so that this phenomenon might be observed at leisure by their guests. Others of those "noble Romans" have been charged with the barbarity of throwing slaves alive into their fish-ponds, with a special view of thus fattening the famous lamprey-eels, which was a fashionable dish of the period.

But the moderns must have the credit not only of cultivating the edible species of fish, mollusks, and crustacea, for useful purposes, but of studying the more minute and curious non-edible, and even microscopic forms, from purely scientific motives. France led the way, and Germany, England, and our own country, have rapidly followed; so that for the last four decades we may say that marine zoölogy has been popularized; and we will yet hope that, in connection with other branches of natural science, it may be made an ordinary branch of education. Why should our youth be kept in ignorance of one-half of the Creator's wonderful works, as displayed in the inhabitants of the waters of the ocean? To prove that the study of natural objects is as good a discipline for the mind as is exclusive devotion to the classics, we have only to point to Cuvier. There is an anecdote told of him which illustrates his own thought on this subject. It is said that, when he was called to fill an important public office, the Emperor Napoleon expressed his surprise at the skill displayed in the management of affairs by this student of Nature; but Cuvier declared that it was precisely the order and system with which Nature requires to be studied that had habituated his mind to grasp at once the problems presented to him in a totally different sphere of thought.

Marine zoölogy has suffered under the disadvantage, beyond other branches of science, that subjects for examination were not easily attainable. During the first century of our era, the greatest naturalist of his day, Pliny, had only discovered forty-seven kinds of marine animals! Successive observers added their mites of observation, but knowledge on this subject was of very slow growth, for, until the invention of the microscope, the numerically largest division of these curious creatures were invisible; and we may say that, until the era of scientific expeditions and the invention of machines for deep-sea dredging, thousands of curious forms of coral, shells, algæ, and all their congeners, were never beheld by any human eyes.

But the human imagination was never idle, though science has been so tardy in its marches; and those poetical fancies, which conceived of mermaids and fairy-grottoes beneath the bright sea-waves, were in many respects nearer to the truth than the so-called facts of some of the old naturalists, for certainly no fairy-land could exceed in beauty many of the gorgeous bowers formed by the combined productions of the marine flora and the animated dwellers in the submarine depths of the tropical seas.

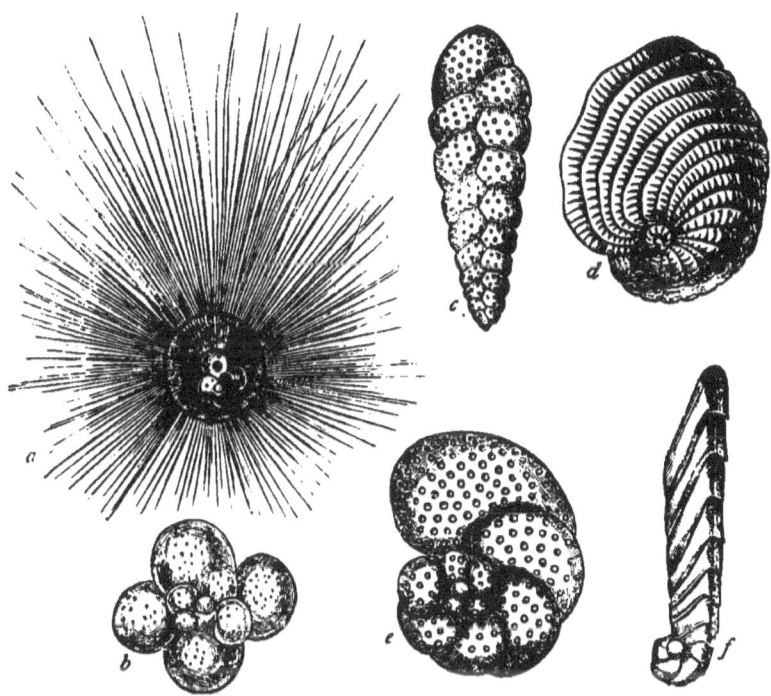

SHELLS OF LIVING FORAMINIFERA.—*a, Orbulina universa,* in its perfect condition, showing the tubular spines which radiate from the surface of the shell; *b, Globigerina bulloides,* in its ordinary condition, the thin hollow spines which are attached to the shell when perfect having been broken off; *c, Textularia variabilis; d, Peneroplis planatus; e, Rotalia concamerata; f, Cristellaria subarcuatula.* Fig. *a* is after Wyville Thomson; the others are after Williamson. All the figures are greatly enlarged (after Nicholson).

In those masses known as *macciotta* an almost unnumbered variety of garden and even forest-like structures are formed, consisting of polypiers, hydroids, corals, algæ, and sea-anemones, of the most brilliant hues and graceful forms, which neither pen nor pencil can ever adequately describe.

It must here be noted that modern facilities of observation have greatly extended the area of animal life ; not only in our recognition of about 500,000 varieties of marine forms, of which 400,000 take rank as very low organizations, but in the discovery that many species of growths once classed as vegetables, such as the sponges and corals, are now known to be animal organisms.

CHAPTER II.

ACTINOIDS.

ACTINIÆ, SEA-ANEMONES, OR SEA-FLOWERS.

HAVING been peculiarly successful in keeping alive and in healthy condition a large number of these beautiful creatures, I shall now attempt to describe a few of them, beginning with a class which is better known than many others, and which must ever grow into popular favor from the beauty of its form, the brilliancy of its colors, and the comparative ease with which they may be preserved in the aquarium. I refer to those charming sea-flowers, the anemones, many of which rival in beauty the choicest treasures of the garden or conservatory. But added to their loveliness of form and color is the superior attraction of their vitality ; for these sea-flowers are living animals, breathing, eating, digesting, and capable of changing their forms at will. Would not a pink be more curious if it could walk ? a rose awaken greater interest if it could reach after its necessary nourishment, and take care of its own buds ? Well, this is what the flowers of the sea do.

This animal-flower is widely diffused, is found upon all our shores, usually adhering to rocks, and has even been found upon the timber of our docks. Every sea offers us some representatives. It is classed by naturalists with the corals ; and, if we cannot all have the latter alive in our aquaria, we can all secure sea-anemones, which will assist us to understand the mode of life of the former.

2

The sea-anemones possess the power of altering their shape to an astonishing degree. Sometimes they will contract themselves into balls, partially elongated and expanded; then they will stretch out their fringes and tentacles to their widest extent, like a polypetalous flower in full bloom.

ANEMONES, OR SEA-FLOWERS.

The variety known as *Metridium marginatum* is particularly fond of this habit. I have long had in my aquarium a beautiful salmon-colored one, which will assume many different forms in as many minutes. Standing up at its full height, some six or seven inches, a constriction or belt will

appear around the middle of its body or column ; this will be
drawn tighter and closer, until a perfect hour-glass form is produced ; then perhaps this band or girdle will move upward
toward the crown, and it will appear like a small umbrella
or mushroom.   Then it may shift downward, and at a certain distance become highly suggestive of a delicate-waisted
young maiden.   I have seen two of these belts appear at the
same time, and have even observed three.   Some of the
forms assumed are wonderfully graceful and interesting to
the observer, but I doubt if these are healthy indications.
Sometimes it remains attached to a piece of rock, and anon
it prefers the smooth glass wall of the tank.   When in the
latter position, the observer is enabled to note the peculiar
structure of the base, and in some of the German varieties
which I possess there is sufficient transparency to give a view
of the internal structure.   In the base of these latter appear
both radiate and concentric lines, indicating the divisions
which exist in the body, forming a succession of chambers ;
a central cavity is observable above and below, forming the
axis of the animal, and around this the chambers or cells are
arranged in a radiate form, char-
acteristic of the class *Radiata.*
These partitions forming the chambers or cells do not all extend to
the inner sac or stomach ; they increase in number in proportion to
the age of the animal.   Around
the edges of these partitions are
the genital organs, arranged close
to the central cavity, in the lower
end of the inner sac, through
which the matured young pass,

CROSS-SECTION OF A POLYP, OR SEA-
ANEMONE, SHOWING THE SEPTA.—
(Dana).

and thence through the mouth opening at the top.
    The upper chambers of the cavity are prolonged at the
superior edge into tentacles or feelers, which extend in a

number of rows around the upper part of the animal, forming when they are all extended a beautiful crown. If these
tentacles or feelers are touched, or if the creature is in any
way alarmed, they are instantly contracted, and all the parts
sink down and are drawn together into a compact mass.
This is effected by the exudation of water from the cavities
or chambers through a series of small openings connected

SEA-ANEMONE AND ITS YOUNG.   ANTHEA CEREUS (Opelet).

with the central cavity. Expansion takes place by the reversed action, namely, filling these cells with water. These
creatures possess in a limited degree the power of locomotion by means of two sets of muscles, one set running around
the body, the others arranged longitudinally; and by contracting and expanding these they are enabled to move even
considerable distances. I have seen them go even at a snail's
pace across the tank.

The sea-anemones are delightfully varied in size, color,
form, and special peculiarities of development and function;
so that a large collection would be like an animated flower-
garden, composed of carnations, china-asters, dahlias, daisies,
etc. Their general resemblance to flowers first caused naturalists to bestow upon the whole class the name of the anemone, or wind-flower; while individual features and peculiarities have induced successive discoverers to name the
varying species in a similar way. Hence we have among
our sea-anemones such designations as the *dahlia, wartlet,
Sagartia rosea, Actinoloba, Dianthus*, and so forth.

The beauty of many species is greatly enhanced by the fact that several colors are combined in individual specimens. Thus sometimes the main body or column will be green, with white or golden tentacles, and the base buff with a pink disk or tips, or crimson with azure spheroids ; sometimes the whole animal will be of one color, varied by different tints and shades. Down below, in the caves of the sea, these wonderful creatures have for untold ages anticipated our modern "combination-suits," and have appeared dressed in all the glory of scarlet and gold, pink and gray, blue and white, green and crimson ; their exquisite taste always selecting accords or pleasing contrasts, and avoiding all discordant shades which would clash with or "kill" each other, such as we sometimes see in human productions.

In my own limited collection there is a beautiful crimson anemone, the *Actinia mesembryanthemum*, a native of the Bermuda seas, with bright blue eyes,[1] situated at the base of the outer row of tentacles ; they are about the size of No. 6 shot, and brilliant as turquoise, to which mineral they bear much resemblance, though brighter. I have others of a greenish gray with white tips, and a splendid specimen from the coast of Massachusetts of a delicate salmon tint, the disk being of a deepened tinge. The same waters furnish other varieties, the *Metridium marginatum* being probably the most numerous. There is also in my aquarium a lovely pure white specimen, taken from New York Harbor. Thus, it may be seen, we have these elegant marine beings at our very doors ; but how few are there to appreciate them as they deserve !

As some of our readers, who reside at a distance from any public aquarium, may never have had the opportunity of examining a living anemone, we will describe, with as

[1] Eyes only in appearance. They are not organs of vision, but a rudimentary form of pigment-cells, such as are found on the margin of the naked-eyed medusæ.

little use of technical language as possible, their general form, habits of life, and modes of reproduction, all of which processes are exceedingly interesting, and may be watched at leisure through the glass walls of the aquarium.

The sea-anemone does not belong to the lowest order of marine life, but takes rank in the third great division or sub-kingdom, the COLENTERATA, sub-class *Actinozoa*. The column-shaped body of this animal-flower is soft, but usually tough and tenacious, and consists of a simple sac, commonly broadened at the base and open at the top or mouth. Within this outer sac is a smaller one, which, though serving for a stomach, is open at the bottom; but its internal walls are provided with numerous *septa* (vertical muscular partitions), which exert a strong contractile force, and prevent the food from escaping until it has performed its office of nourishment. This sac the animal can turn inside out without injury or inconvenience. The upper edges of these two sacs are united, forming a thick circular lip, from which proceed the numerous tentacles that fringe the mouth with a collarette of sensitive and retractile projections, individually closely resembling the petals of certain polypetalous flowers; but they are narrow in proportion to their length, and, instead of being flat like most flower-petals, are hollow and susceptible of extension to the point of transparency. With them they clasp their food, and perform other curious functions to be hereafter mentioned.

In addition to the tentacles, these curious creatures are armed for attacking their prey with what we may call fine thread-like lassos, of arrow-like sharpness, called *cnidæ* (from a Greek word meaning a nettle), from which is transmitted a powerful stinging and benumbing sensation, deadly to small prey, the victim being affected as by a shock of electricity. This I know by experience, for, some years ago, when in Bermuda, while attempting to take a large actinia from a rock, one of these soft-looking beauties gave me a shock which disabled my arm for hours.

It will easily be understood that this concealed battery enables the sea-anemones to conquer much larger and stronger creatures than they could hold simply by the tentacles; they often seize large shrimps, and crabs far beyond their own size. Occasionally, however, if one of these finds an anemone weakened from any cause, it will take up a position upon the edge of its mouth, keeping it distended, and with its claws pluck out the food from the victim's sac and appropriate it to its own use. Sometimes, when such an attempt is made, a combat ensues, and then woe to the marauder if he has mistaken the strength of the sea-anemone! He will surely fall into his own trap.

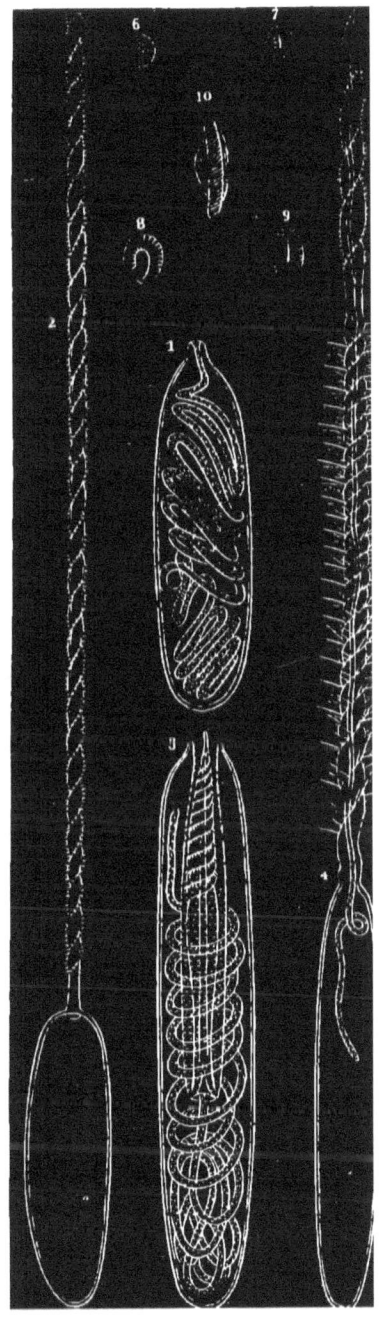

Some of the sea-anemones are free-swimming organizations, and all have some capacity for movement; but the
habit of most is to attach themselves to some firm object, as
a rock or a section of coral, or to the back of a crab or other
crustacea. In fact, when free they swim backward, and
wherever their base encounters a firm object, no matter
what, there they will fix themselves by suction, and as a general rule contentedly remain. There are two species, however, which show a marked preference for the back of a crustacean. One is called the parasite anemone, and its favorite
home is on the hard shell of the hermit-crab (the *Pagurus*

HERMIT-CRAB WITH SEA-ANEMONE ON ITS SHELL.

Bernhardus); and as these crabs are great travelers, and have
the peculiarity of frequently changing their residence by
taking possession of the empty shells of other animals, this
parasite anemone is likely to see far more of the world than
its more modest brethren. There is one other genus which
cultivates the parasitic habit, the *Adamsia*, which selects the
crab *P. Prideauxii* for its place of abode. This habit is
known as commensalism, as they are presumed to dine at the
same table.

Once located, the sea-anemone has only to continue sedentary, to open its mouth, and wait for food to float within reach of its tentacles, and ocean-water furnishes a never-failing supply of the lower forms of infusoria, zoöphytes, and polyps. I give mine small portions of clams, oysters, or sometimes scraped chicken. If anything enters their mouth which proves unacceptable, they very promptly eject it.

Anemones may live singly or in society, but they readily tend to reproduction, as most of them are hermaphrodite and are singly capable of producing living germs. Besides, the anemones are not reduced to a single mode of reproduction; and though the birth of an anemone might in the aristocratic circles of the *Actinozoa* be considered more strictly legitimate when the germ is nourished within the sac, yet this mode is not uniformly adhered to. When, however, this is the case, the larva remains within the sac until a certain stage of development is reached, when it is ejected from the mouth of the parent, and sets out on its own account, a free though as yet an imperfectly developed anemone—needing only a little more time to complete its structure, affix itself to a permanent base, and become the nucleus of a new society. When these young are first set free, they usually appear with only six tentacles; subsequently these are multiplied by sixes to twelve, twenty-four, forty-eight, and so on, until in the full maturity of the creature they sometimes approach to two hundred in number.

Nearly or quite all scientific writers on zoölogy represent this as the *invariable* procedure in this kind of propagation. But one of my anemones has, within a few weeks, given birth to seventeen young; and certainly in two instances which I observed, instead of ignominiously ejecting them from the mouth as above described, these *were tenderly taken from the mouth by two tentacles, which then became wonderfully elongated, and these young larvæ were carefully deposited on the rock which served the mother-anemone as a base.* They are now alive and wonderfully active in their independent existence. To the uninstructed eye it was as if one looking at a dahlia or aster should suddenly see the flower pluck a bud from its centre by two of its petals, extend these to the earth, and there deposit the germ, which should thereupon begin to develop into a mature flower! Would not such a sight be worth going a long journey to see? Yet as curious actions, by these and other small marine animals, may be witnessed by any one sufficiently interested to keep and faithfully watch even a small aquarium.

But this internal process of germination is supplemented by two other kinds : *gemmation,* or budding, and *fission,* or division. In the former the young are formed on the outside of the parent, on the side or near the base, from which at the proper moment they detach themselves and become independent existences. By the process of *fission,* the severed parts, as with many polyps, grow into perfect individuals. Should an anemone be divided horizontally, the lower part would soon form a new mouth and tentacles; the upper part would remain for some time open at both ends, fruitlessly endeavoring to satisfy its appetite for food which floated through it — a more hopeless effort than that of Tantalus —but Nature, kinder to the anemone than to him, soon comes to its relief, as the lower portion eventually closes, developing a new sac and tentacles, and the creature continues to live as a twin or double anemone. If divided vertically, there is a tendency to reunite and form two parallel growths, each somewhat slimmer than the original.

Some species have a habit of throwing off portions of their base, which form into new individuals. The *Actinia dianthus* has this *fissiparous* habit, and others may suffer vivisection from man with similar results. The severed portion may be without germs or ova; it is only necessary that the piece cut away should contain the three elementary tissues of the animal, i. e., the *tegumentary,* the *muscular,* and the *ciliated lining membrane,* so tenacious of life and full of vivifying power are these delicate-looking creatures.

It would take a volume to fully describe even a small portion of the numerous varieties of sea-anemones.   But, among those which have been most successfully preserved in aquaria, we may mention the parasite anemone, which affects the hermit-crab; and this, we may say, has been proved to be a deliberate and not accidental partnership. Gosse relates that, having found an anemone on a shell de-

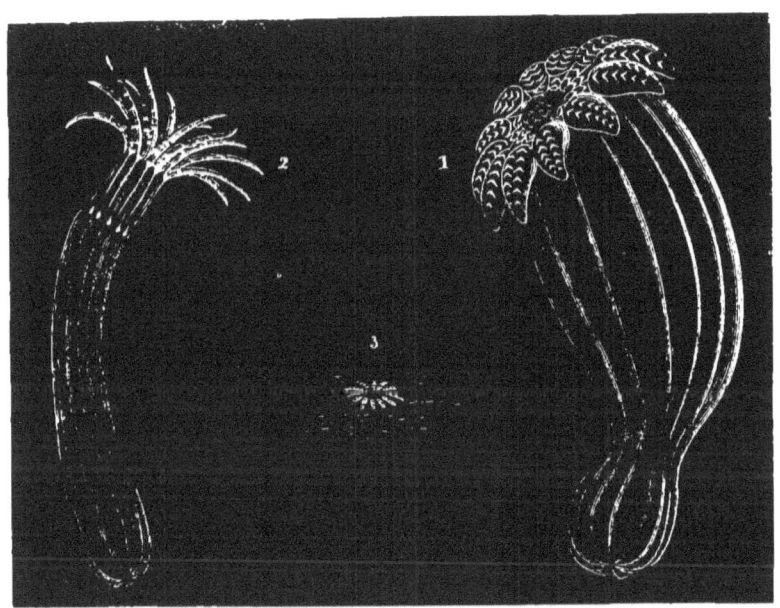

ACTINIÆ, OR SEA-ANEMONES, WHICH LIVE IN THE SAND AND ARE OFTEN UNATTACHED.—
1. *Peachia hastata,* Gosse.—2. *Edwardsia callimorphia,* Gosse.—3. *Halcampa chrysanthellum,* Gosse—the last mostly buried in the sand.

serted by a hermit-crab, he put it, still adhering to its old home, into his aquarium.   It soon left the deserted abode, and stuck to the glass walls of its prison; but the crab being placed in the tank, and again taking possession of the shell, the anemone left the glass and returned to the back of the crab and remained there.   The cloak-anemone always seeks to base itself on the inner lip of some univalve shell. The sand-pintlet, *Halocampa chrysanthellum*, needs only sand enough at the bottom of the tank to burrow in ; and, when he raises his beautiful head and looks about for food, he easily pays for his lodging, which is certainly inexpensive.

The globe-horn, *Corynactis viridis*, is of a beautiful yellowish emerald-green, sometimes of a translucent white, having umber tentacles, or brown with pink tips. This species has the peculiarity of not embracing its food with its tentacles ; it simply opens its mouth wide and receives what floats toward it.

A curious species is the vestlet, *Cerianthus Loydii*, about seven inches long. It has in its natural state a rough, felty coat, which can be stripped off without injury to the creature, and to prevent its reforming it may be kept in a glass tube within the aquarium.

Some of these sea-anemones secrete calcareous matter or corallum, and are considered a connecting link with the true corals. Of this class are the *Capneadæ*, of the tribe of *Caryophylliacæ*.

The eyed *Sagartia* is so named from the fact that when its tentacles shrink, or are withdrawn, there remain small, elevated points resembling the eyes of a butterfly's wing.

Some dwellers by the sea in the south of Europe, more blessed with good appetites than æsthetic taste, do not hesitate to cook and eat these beautiful sea-flowers ; the taste is said to resemble that of the soft crab. The celebrated English naturalist Gosse also tasted them, and pronounced them superior in flavor to the periwinkle.

> "Full many a gem of purest ray serene
> The dark, unfathomed caves of ocean bear ;
> Full many a flower is born to blush unseen,
> And waste its sweetness on the desert air."

CHAPTER III.

" We wandered where the dreamy palm
Murmured above the sleeping wave,
And through the waters clear and calm
Looked down into the coral cave."

AMONG the advantages of travel may well be reckoned the memories of scenes passed through—the adventures and labors shared in common with sympathizing companions—especially when the object of the journey was the observation and study of natural productions, fauna or flora, on land, river, or sea. Every practical marine zoölogist must have shared in the keen delight and curious expectancy of watching the rise of the dredging-machine, as it approached the deck from its foraging excursion below. How we hoped to find this or that—some particular specimen upon which we had set our hearts; and with what disappointment we turned away if nothing of value was discovered, or only the commonest specimens appeared, of which we already had abundance! But did the eye perceive some unknown form, with what ardor it was secured, and yet with what gentleness and delicacy it was handled and inspected!—for experience had taught us that some of the most beautiful marine forms are not to be touched with impunity, many of them possessing stinging qualities, while others, like the brittle-star, have the inconvenient habit of dismembering themselves if displeased or frightened.

There are many favorable locations for finding varied and curious specimens, such as the waters of the Mediterranean, the shores of Japan, and the coral islands of the Pacific; but, for those who cannot make extended voyages, there is perhaps no better hunting-ground for marine curiosities than the Bermuda islands. One reason for this probably is, that in favorable years the Gulf Stream throws many exotics on its shores; but a more permanent cause may be found in the fact that this group of islands is entirely organic, and that both fossil and living specimens of corals, mollusca, annelids, and wondrously beautiful fishes, may be found in abundance. But it is of coral alone that we now design to speak, and this interesting production may also be sought among the Florida Keys.

It is not strange that so curious and beautiful an object as coral should have early attracted the attention both of naturalists and ordinary travelers. Even the common seaman likes to take home a piece of coral to adorn his humble abode, while learned scientists have reasoned and argued with pertinacity and zeal over the mystery of its construction. The Greeks named it the "Daughter of the Sea," but are not known to have investigated its nature or mode of growth. For a long period it was the subject of curious conjectures, such as that it was a vegetable formation, and again that it was soft while in the water, and only hardened on exposure to the air; and even to the present time there remain in the popular conception several curious errors in regard to its growth. Indeed, we have heard public speakers, clergymen and others, in pursuit of an illustration, speak of "the wonderful *labors* of the coral *insect !*" In this short phrase are involved two fundamental errors; for the coral-producers are neither laborers nor insects.

Another very common mistake is the supposition that they are exceedingly minute — even microscopic — in size. This is far from being the case. Having had several varieties under observation in my aquarium for years, I can assure the reader that they are not only large enough to be plainly seen by the naked eye, but that they sometimes elongate themselves nearly an inch above the upper edge of their cell, measuring one-third of an inch in diameter.

But some one may ask, "If the coral-producers are not insects, what are they?" We answer, mainly polyps, with some hydroids and soft mollusks of the lowest class. These are all soft-bodied organisms, consisting of many varieties, having the organic function of secreting carbonate of lime, which, with some other ingredients, as silica and small portions of sand, composes the hard substance called coral.

The body of the polyp consists of a cylindrical skin, with an inside sac, which is the stomach, and is furnished at the top with thread-like appendages, with which it draws in its food. Whatever it does not wish to retain in the stomach it rejects by the mouth, having no other resource, as the lower end of the polyp is affixed to the stony substance. When expanded, these thread-like tentacles around the mouth give them a flower-like appearance. It is between the outer skin and the sac or stomach that the limestone is secreted which forms the coral substance.

It will thus be seen that the polyp does not gather or collect from external sources the material of the coral—does not in any correct sense work or "build" any more than a tree may be said to work as it grows into wood. Nature has simply provided that, in receiving its food, the polyp selects from the ingredients of the sea-water that which is capable of being reduced by simple functional processes into coral; just as a plant selects and secretes from the earth that kind of nourishment which makes stems, leaves, and buds.

Each mature polyp, when fixed in its cell, may be considered as resting upon the tombs of its ancestors; and, when it dies, its descendants will repeat the process over its remains, and its own body, within which its share of coral has been secreted, will be the base for a new living descendant.

The cells of the coral colony are not retreats into which the polyps come and go, like a bird building its nest, but part and parcel of the creature; just as much so as are the bones belonging to a human body—with this difference, that in the case of the polyp the stony part (representing the osseous structure in man) is all at the lower portion, while the upper part is soft and flexible; but in a healthy condition they are inseparably combined.

It is well known that the power of secretion is inherent in all living tissue, while its matter and form are varied in every possible degree, from the animalcule to the superior mammal. This power or faculty is possessed in full perfec-

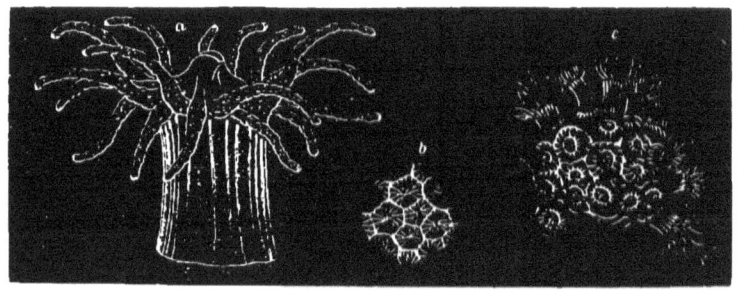

DANA'S ASTRANGIA (*Astrangia Danæ*, Agassiz): *c*, a growing cluster; *a*, a single polyp enlarged; *b*, the dead coral.

tion by what we call the lower forms of life, and it is among these we find the stone-makers; for the simplicity of their structure is such that they may be nearly all stone while yet the small portion of vital substance carries on the processes of nutrition and growth.

It is not generally known, outside of the circle of naturalists and the scientists connected with the Coast Survey service, that coral grows on our own North Atlantic shores, the popular belief being that all corals are to be sought for in warm climates. There is one variety at least, *Astrangia*

Danæ, which has been found on the shores of Massachusetts and Connecticut. But what we call true coral has not hitherto been kept in a living condition in any other private aquarium than my own. So far as I know, my acquisition is unique, being the *only living specimen of the true coral-producing polyp preserved in captivity within the United States,* though plenty of it may be found at the Florida Keys, some parts of the West Indies, and at the Bermudas. But these delicate creatures cannot be preserved except under favorable conditions; and constant care is required when in the aquarium to see that no injurious substance or fellow-captives come in contact with them.

The *Astrangia,* specimens of which have been found in the vicinity of New Haven and other points along the Sound, differs in this particular from other varieties of coral: the polyp rises more prominently above its cell; the coral secretion being limited to its base, and not, as in the reef-building and some others, continuing the secretion between the outer walls of the internal sac and the epithecum. The tentacles have minute warty prominences upon them, which are full of lasso-cells. I have never seen specimens of this variety above three or four inches in diameter, by less than one in height. The specimens I have are fine indeed. I have seen them when the mass of little animals with their myriads of fleecy locks looked like a pure white ball of snow.

Among other varieties of coral which I have succeeded in preserving in health and good working order is a fine specimen of the species known as *Occulina.* Its peculiarity is its branched or tree-like form; its zoöthome, or mass of living occupants, when *out in all their glory,* are so numerous as nearly to hide the stony substance of the corallum, presenting the appearance of a stem and branches adorned with *living flowers.* In this variety the tentacles are shorter than in some other kinds, nor have they the power of elongating themselves to the same extent. They are colored, but

not uniformly of one shade; they are often greenish or of
an umber tint, with changeable *iridescent* color. They are
usually considered more elegant than the reef-building variety, from which their internal structure does not materially differ.

It will be perceived that the association of the polyps in
all compound groups must be of the most intimate kind;
for, though each individual has a separate mouth, tentacles,
and stomach, yet the intervening tissue which connects them
is subject to a free circulation of fluids through its pores or
lacunes. Thus the zoöthome as a whole must be considered
as a living mass of animal matter, which is all benefited by
the nourishment received through the individual members.
It is a perfect commonwealth of its kind, even exceeding in
perfection that of the honey-bees; for there are no drones
among the polyps.

How long the coral polyps will live is not known. I
have preserved mine in pure sea-water for years, and they
appear to be as fresh and healthy as ever, active and industrious; but this industry, as has been explained, does not
consist in any such operation as "building." Their simple
and sole business is *eating;* and that a strong stony structure
is the result, is no more creditable to them than it is to a
maple-tree to secrete sugar, nor does it indicate any more
effort.

The process of coral growth is, however, very slow, if my
specimens are any criterion. But a very minute addition
has been made to my corallum during the years that I have
had it. I cannot say, however, whether it might not have
progressed more rapidly had it been left in its original
habitat.

But though my corallum has grown so slowly in height
or lateral extent, it has increased very fairly by gemmation
or budding; a considerable number of infant polyps have
been added to the group, always, so far as I have observed,

in the autumn, October and November. At first the young larvæ are worm-like in form, whitish and semi-transparent; they are very agile, and dart about in all directions, swimming, as we may say, "stern foremost," as their mouths are always in the rear. But this life of freedom soon comes to an end : Fate has ordained that they shall become fixed to their parent-stem or some other stationary object. Their mode of swimming facilitates this result, the base having a natural tendency to adhere on contact; and thus its gay youth is soon exchanged for a sedentary life, with no other changes than that of eating and digesting their food.

There are few natural objects more pleasing than an association of these corallets ; for, as the polyps rise above their cells and extend their fine long tentacles, resembling threads of pure white silk, waving them to and fro like the radiated petals of a fairy-flower swayed by a gentle zephyr, or, again, like a minute feather fan slightly concave at the edge, they present an exceedingly animated and elegant appearance. Sometimes, when nearly at rest and the filaments are more contracted, they suggest the appearance of a dense frost settled upon a bed of moss.

But these fairy-like implements, with which the coral-polyp gathers in its food, are not such innocent objects as they appear to the naked eye. Examined under a magnifying-glass, there may be observed on the tentacles a row of slight protuberances, one larger than the rest being situated at the tip. These might easily be mistaken for ornaments, but their character is far different : they may, in fact, be looked upon as the creature's armory, for within them lie concealed *cnidæ* or lasso-like filaments, sometimes called capsule-threads, which are capable of being thrown out to a distance many times the length of their own bodies. It is with these that they capture their prey ; for these little soft animals are carnivorous in their habits, and indeed have not yet abandoned the barbarous practice of infusing poison into their darts.

In each of these *cnidæ* is secreted an injurious fluid, which
partially or wholly paralyzes the small crustacea or other
animals which the tentacles seize ; and whatever small prey
falls within their grasp is very promptly and certainly secured. The mouth of the polyp being in the centre of the
upper portion of the body, the victim thus seized is rapidly
passed into it by aid of the longer tentacles, and thence descends to the stomach.

The *actinoids*, to which class these coral-polyps belong,
owe nearly their whole success in foraging to these concealed
weapons, which are numbered by myriads. The cavities
which contain them have been called by Agassiz lasso or
nettling cells. Gosse names them *cnidæ* or thread-capsules,
because the small cell-shaped sheath contains these slender
tubular threads, coiled up, ready for use ; and they are darted
out with astonishing rapidity when a victim happens to stray
within reach. The poison is communicated instantly at the
first contact.

Should you have the least feeling of doubt as to the
stinging propensity of these little animals, you may readily
satisfy yourself on the subject by just touching your tongue
or lips to the surface of the corallum, when you will receive
such a *sensational explanation* on this point as you will
never forget. In handling and arranging living corals from
time to time, I have felt this poisonous stinging sensation,
and suffered considerable pain for an hour, and some degree
of inconvenience for more than twenty hours after. The
shock from these, however, is much less severe than from
the anemones, as mentioned in the chapter on those living
" sea-flowers."

The food received into the stomach is always digested
before the animal retires to its quiescent condition within its
cell or *calicle*. If closely observed, this process can be actually seen through the semi-transparent, glassy walls of the
living polyp. I have tested this by furnishing my pets with

small portions of our ordinary edible mollusks, not over the fifteenth of an inch in size—perhaps as large as half a medium-sized Zante currant.

In the *Corallidæ* the axis is wholly calcareous, firm and solid throughout, of a color usually varying from crimson to rose-red. In this class belongs the *Corallium rubrum*, the red or precious coral. The cortex—that is, the outer crust, resembling somewhat the bark of a tree—is in this species thin, contains comparatively few calcareous spiculæ, and may be readily rubbed off when dried specimens are handled. This outside cortex is of a coralline nature, but the true coral is the red axis which it envelops. When specimens of this variety have been carefully preserved, the polyp-centres may be perceived; they are distinguished by the appearance of a faint six-rayed star. When living they are similar to other alcyonoids in respect to the number of their tentacles, there being eight, and fully fringed. In the living corallum they open out profusely, as was described in the branched form, making an exceedingly beautiful object. The branches have a tendency to expand horizontally.

Although I have not yet been fortunate enough to obtain a living specimen of this variety, I hope soon to receive one, and may yet be able to describe their habits and growth from my own observation. Large quantities of this precious coral, or coral of commerce, are annually fished up from the rocky bottom of the Mediterranean Sea and carried to the Paris and London markets, whence much of it reaches New York. Some extensive dealers, such as Messrs. Tiffany & Co., of this city, use in their jewelry hundreds of thousands of dollars' worth annually. Some shades of it are immensely valuable. The *real rose-pink* is so much sought after, that its value becomes almost fabulous, being sometimes sold at *twenty-five times its own weight in pure gold;* it is always sold by the ounce or pennyweight. When more than a few inches in length, its value is still further augmented, as few pieces of any considerable size are ever offered for sale. This scarcity tends to enhance its worth, as well as the intrinsic beauty of its peculiar color. Mr. Gideon F. T. Reed, of Paris, some years ago kindly presented Mrs. Damon with what might be considered a giant specimen of this coral, measuring some twelve inches in surface; it is likewise of a most graceful branching form, something like the antler of a deer. This she only allows me to *look at* on rare occasions! It is certainly the finest specimen of this coral that I have ever seen. The *Corallium rubrum* is found at depths varying from twenty-five to a thousand feet.

The black coral, *Corallium nigrescens*, is still more scarce, and, though capable of being worked into *bijouterie*, on account of the small number of specimens which have been secured it is very rarely met with in commerce. One handsome variety of this color is the *Antipathes subpinnata*, so named on account of its spiny habit of growth. "It is," says M. Moquin-Tandon, "a fragile and brittle *polypier*. When dry, the branches, slender and delicate, resemble the barbs of a feather." It is sometimes found with a brownish or greenish tinge. The bark or cortex is soft and pliable, being destitute of calcareous or silicious matter, and is easily rubbed off when the coral is dead. The polyps are of a yellowish color, and long as compared with other varieties.

The large, massive forms of coral, whether of the dome, reef, or tree-like shape, would never reach the magnificent proportions that they do were it not for that peculiar provision of Nature in regard to the zoöphytes, of life and death both proceeding simultaneously and successively; each, combined and singly, aiding in one and the same object. This curious condition of growth favors the coral aggregation by allowing the living polyp, as it secretes the calcareous matter, to mount upward on that which it has already secreted and deposited. From the successful execution of this ascending process, we are led to infer either that the creature has the power of indefinite elongation, or that it must desert the precipitated portion of the corallum as growth proceeds; and, in fact, this last is what actually occurs. In some instances a polyp of only an inch in length, and even less, has been found at the top of a stem many inches in height; for the whole substance of what is called "living coral" is in reality dead, excepting the extreme surface or point of each branch occupied by the little animal. The living tissues which once filled the cells of the lower portion of the corallum have been consumed by natural processes, and have disappeared as growth went on above. Some writers, speaking of this subject, use the expression that these tissues have "dried away;" but how they could "dry" away under water is not very clear.

On this theory of the growth of corallum, it would seem that there is no necessary limit to its increase in height, and that the elevation might extend indefinitely; but, practically, we find that its natural limit is the surface of the sea. When that point is reached, the polyp dies, apparently from the exposure only, and not from any inherent cause of death in the animal organization. Some one or two exceptions to this rule have been reported, as madrepores having been found "alive and well" six inches above low tide; but the narrator does not say how long they remained alive! In those dense, weighty, dome-like forms called *Astræa*, the rule certainly holds good: when the summit attains the level of the sea, progress ceases as to height, but it may yet be enlarged considerably as to breadth; and, should we endeavor to separate one of these masses into their respective living and dead portions, we should have in the former a hemispherical shell, perhaps not more than half an inch thick, while the dome would be a mere dead mass of stone, perhaps ten or fifteen feet in diameter.

The final solidification of the coral mass is aided by the increased secretion by the polyp shortly before its death, filling all the pores with this stony matter in proportion as the vital tissues occupying them shrink and dwindle. This last deposit greatly aids in strengthening those tree-like or

CLUSTER OF CORAL-POLYPS (*Asteroides calycularis*, Milne-Edwards), in various stages of expansion.

DEAD CORAL (*Asteroides caly- cularis*, Milne-Edwards).

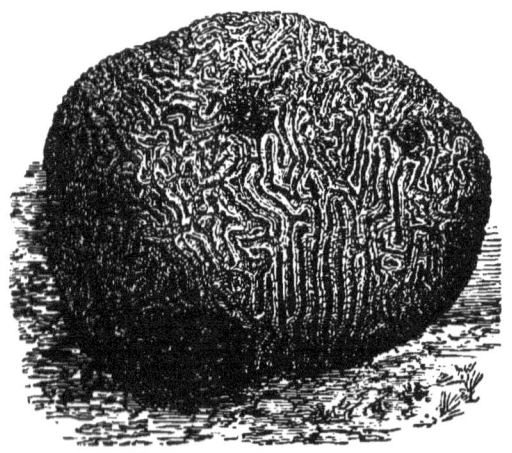

MEANDRINA CEREBRIFORMIS.

branched coral growths which, though so slender of form, are really very strong.

The facility with which polyps repair an injury, a fract- ure, or disturbance of any kind, is common to many of the lower marine organisms. A blow or sudden shock given to a mass of coral would probably cause the whole community to withdraw into their cells for a short time, perhaps an hour or more. Should even a portion be broken off, the inhabitants of the remaining mass would not long remain in retreat; but very soon we should see them, with their upper surface exposed, and their tentacles all expanded, commencing the process of restoration without loss of time. The separated portion would either adhere in some crook or cranny of the mass, and reunite with its old associates, or, if too far removed for contact, would become fixed on some rock or other firm substance, and continue to grow regardless of the accident, perhaps becoming the nucleus of a mass as large as its parent group. It would do this by cementing its base, through new coral secretions, to whatever substance it came in contact with, if in any way favorable to its habits and nature.

The common surf-corals, of which branching madrepore and the dome-shaped astræa are good examples, consist al-

ASTRÆA PALLIDA (living).

most wholly of carbonate of lime—the same ingredient which constitutes ordinary limestone. In 100 parts, 95 to 98 are of this substance, and of the remainder there are about four parts of organic matter, with some earthy ingredients and traces of silica, and in a few cases fluorine has been detected.

3

Its density exceeds that of common limestone or marble, as may be tested by striking it with an iron instrument.

The varieties of coral are very numerous, some growing in masses and others living an independent individual existence, and all very interesting in their way; but, as this little work is designed to give the results of our own observation and experience, we shall not indulge in any extensive description of other varieties than such as we are personally acquainted with. Our only regret, in closing this chapter, is that we feel the utter inability to express by words the charm which these interesting creatures are capable of exerting over the mind of the lover of Nature, who watches them daily, feeds them, observes their changes, recognizes the condition of their health by the greater or less force with which they wave their little fairy flags of white, and follows with the keenest interest the fate and fortunes of the infant polyps, born under the loving and gentle captivity of our crystal-walled aquarium.

CHAPTER IV.

HOUSE-BUILDERS OF THE SEA.

OCEAN architects, house - builders, decorators, masons, weavers, jewelers, dyers, butchers, mowers, scavengers, surgeons! Who would expect to find the trades and professions represented to this extent among the denizens of the ocean? And yet in this list I have named but a portion of the curious marine operators which we have under daily observation in my own aquarium.

It would almost seem that the most ingenious architects and the finest constructive instinct were to be found apparently in the most insignificant forms. It is in the lower orders of animal life peculiarly that there are hidden from common observation worlds of beauty and wonderful displays of skill, of which the transient or careless observer has no conception.

Peering through the transparent walls of my aquarium, filled with clear, bright sea-water, I have for years daily watched these marvelous operations with never-wearying interest. The unexpected revelations of curious instinct, which in some cases seems to approach the verge of reason, the growth, the changes, the curious and beautiful combinations of form and color, are revelations more like a glimpse into fairy-land, or the realization of some fantasy of the imagination, than simple facts in Nature's inexhaustible storehouse. And when we consider how many extinct myriads of these creatures have gone on unnoticed, even unknown to man, for so many ages, exercising skill and apparent taste, with no

eye to observe them superior to their own class, it gives
something of a shock to the pride of man, who is apt to
consider that all the lower animals were made for *his* convenience.

ANNELIDA.

Serpula.—The architectural productions of some of these
submarine workers are of no mean order, each building its
own residence, with separate stones, grains of sand, little
hard pellets of any kind which will answer the purpose—
laying them on in regular tiers as neatly as a mason. Among
these pretty and interesting animals is the *Serpula contortu-
plicata*, which is distinguished from the *Serpula vermicu-
laris* by always holding its head up above the latter, the
upper portion of its tube being nearly vertical. When the
animal is at work, with its feathery plumes fully expanded,
it far surpasses any flower in the extreme delicacy of its
beauty. An inexperienced person might naturally take these
serpulæ for coral-builders, seeing that their stone-like house
is composed of very similar-looking material; but in the
classification of the zoölogists they are entirely disconnected.
They belong to the class known as *Annelida*—that is, worm-
like, as the animal itself really appears when out of its sheath-
like house; but never was worm of the earth adorned with
such a beautiful head-dress as this annelid of the sea. They
are generally found in clusters on the back of a broad shell,
like a mass of tubes, contorted into various curves and twists
at the lower ends, sometimes lying nearly horizontally on the
shell or stone to which they are attached, and then, after
sundry windings and twistings, shooting up in a nearly vertical direction several inches. The beautiful plume or head-
dress, as we have called it, which is the most attractive point
in this creature, really consists of its breathing-organs—the
branchiæ or lungs. Just think of carrying one's lungs on
the top of the head in the form of delicate-sprayed plumes!
The color of these branchiæ is extremely variable; they

number about eighteen separate barbs on each side, sometimes of a purplish brown, transversely marked with flake-
white, intermingled with yellowish green; the pinnæ have
about the same colors. Sometimes we have the whole plume
an orange-brown, white, salmon, chestnut, orange, or an um-
ber-brown.

They construct their tube-like homes from lime, which
always exists in a state of solution in sea-water, and which

AN ANNELIDAN CITY.

the serpulæ know perfectly well how to appropriate and
apply. Mine have added half an inch to the height of their
houses in a few weeks after transferring them from their
ocean habitat to my glass vase. If their tubes are accidentally broken, they are equally apt at repairing, and the deficiency or injury is soon made good. A year ago my
group of serpulæ was tumbled down against the side of the
glass, and a few days later I was very much interested in see-

ing that the industrious little creatures had begun to recon-
struct their broken houses. This, to my great satisfaction,
they did by using the side of the glass tank as a support or
buttress; they thus economized labor and material, while I
had the benefit of being able to watch the whole proceedings
without any intervening obstruction.   This was a fine chance
for learning the secret of these mysterious marine masons, and
their *modus operandi* soon became familiarly known to me.

In addition to the ornamental branchiæ which we have
described, these animals are furnished with a curious appendage which answers the purpose of a front-door to their dwelling; it is called the *operculum*, and looks when shut like a
small disk which completely closes the entrance to the tube,
like a cork in a bottle, supposing that the cork, instead of
protruding, was sunken a short distance below the mouth.
When the creature is frightened, or for any cause retreats
into its shelly tube, this operculum is the last portion drawn
in, and is the first protruded when the owner chooses to
reappear.   Its movements in the act of withdrawal are so
rapid that I have never yet actually seen it shut its door,
but on examination have found it securely closed to all in-
truders.

When additions of any kind are made to the aquarium, the serpulæ exhibit great timidity; even the passing
of my hand over the glass will cause a retreat, or the sudden shifting or raising of a curtain which affects the light
thrown upon them.   Equally on a changeful day, when sun
and cloud alternate in the sky, these atmospheric changes
affect the movements of these sensitive creatures.   From
this extreme sensibility to movements and shadows, it is reasonable to infer that the creature has organs of vision, though
naturalists have not yet succeeded in pointing out the location of the eyes.[1]   Familiarity and habit soon reconcile them

---

[1] M. A. Quatrefages thinks he has discovered eyes *upon the branchiæ* of the
Sabella, an annelid closely related to the *Serpula*.

to captivity, and in a few weeks they become quite domestic, and even sociable, so that they can be fed; and, if they chance to be inside when I approach, *they will come out to see me.*

There are several varieties of serpulæ. That kind known as the *Stillata* forms its valve or operculum of three distinct plates, threaded together, thus arming itself with a triple door. Another variety, called *Citrina*, builds its tube inde-pendently and lives alone, as if it objected to neighbors, while most of the other varieties are found in groups; it is of a bright lemon-color, and very beautiful. Its operculum is somewhat shorter and thicker than that of the variety first described.

The apparatus by means of which the serpulæ perform their upward and downward movements is a marvel of inge-nuity. The body of the annelid is composed of seven dis-tinct segments, and from each of these projects a pair of tubercles, each containing a bundle of bristles, which can be thrust out at the will of the animal; at the end of each of these bristles are four sharp points, one being longer than the others. In ascending, these bristles are thrust against the walls of the tube, which gives the creature a forward impulse; contraction follows, the hinder set of feet-like bris-tles are brought up, and so the movement is repeated till the end is accomplished. Now for the descent: Attached to each of these bristle-like feet is a small ring, resting on tri-angular plates; each plate is notched with seven teeth; six turn one way, and the seventh is reversed. There being one hundred and thirty-six on each ring, and as many rings as feet, there are fourteen times one hundred and thirty-six of these prehensile plates. With all of these teeth the ser-pula can seize upon the lining membrane of its tube to aid in its descent..

Spirorbis.—How many of our readers have picked up on the sea-shore specimens of sea-weed on which they observed small white rings which looked like a coralline substance?

And how large a proportion of those who have looked for a moment at the little circles have realized that they held in their hands a whole colony of living or lately living creatures? These are the *Spirorbis*, a minute variety of annelids, and nearly related to the showy serpula. It is usually found attached to the leaves of the *Rhodymenia* or the *Chondrus crispus*, though it is also found upon rocks and stones. It is usually found in considerable masses on the flat surfaces of the leaves or stones. The shells are very small indeed, looking something like the "mite sugar-plums" or small coriander-seeds; the shape is spiral, and it generally consists of only one whorl. These little creatures cement themselves so firmly to the leaves that it would be impossible to shake them off. From the largest end of the little tube-like circle they put out lovely infinitesimal, plume-like branchiæ, resembling those of their larger relatives the serpulæ, but so fine and delicate that it requires a lens to bring out all their beauties. In almost every collection of algæ will some of these tiny shells be found adherent on the preserved specimens, but to secure them living is not easy. It was not until about a year ago that I was able to add a live colony of these exquisite little spirorbes to my aquarium.

Sabella.—Among the groups of my serpulæ another family has found good quarters; they are evidently relations, and apparently on good terms, though not following the same fashions exactly. These are the *Sabella* tribe, inhabiting tubes like their neighbors, but made of different material. Instead of being formed of hard limestone, the tube is integumentary, or of a leather-like texture; its diameter is about the same as that of the serpulæ, but it generally rises somewhat higher. From the upper orifice of its house or tube the sabella displays a much larger wreath of plume-like appendages, which in its graceful form rivals its neighbor the serpula. Its color is not so variegated or brilliant, but it makes up for this deficiency by its superior size; the disk in some of my specimens would measure over half an inch across.

This little creature has also its means of locking up its house, and can conceal itself within as securely as its relatives in their harder shell; and, when in danger from the hungry tramps of the sea, it can drop down as promptly, and close its door in the face of all obtruders, with full confidence that none will dare to molest it.

Another curious little animal, also an annelid, is generally to be found rearing its cozy home amid the tubes of the serpulæ. It cannot boast perhaps of as much beauty of color and waving plume, but its habits are so interesting and really wonderful, that I think it takes the lead as an object of curiosity of all the tube-building fraternity. Its tube is not homogeneous in its composition like that of the serpula, but it makes an aggregation of separate particles, artistically welded or fitted together like a piece of mosaic-work. This tube is not a secretion, like the cell of the coral-polyps; it does not *grow*, but is voluntarily and with great skill and care built up by the animal. In its construction it will use the very finest material—little specks of fine sand, and even dust that may chance to fall on the surface of the water. It also discriminates as to color, apparently preferring the brighter particles. For instance, I have ground red coral to powder, and put it into the water; upon this the little annelid would promptly seize, and immediately appropriate it for building purposes. Sometimes its tube presents not only a showy but thoroughly patriotic appearance, displaying the national colors of red, white, and blue! In size it varies considerably, averaging somewhat less than that of the serpula.

A casual observer might see this wonderful worker many times without perceiving or appreciating its artistic movements; but get him once under a good lens, and you will see not only all the machinery in full operation but also the

object of its unwearied toil. Indeed, this busy little work-
man lifts and carries bits of stone (hypothetical bricks), grains
of sand, coral, glass, or shell, or any atoms which will serve
its purpose, raises them to the top of its unfinished walls,
and there places them with as much precision, neatness, rapidity, and in as regular order, as the most experienced brick-
layer. It is perfectly marvelous. One might watch them
for hours together and never grow weary.

But *how* do they do it?

When the operation is seen, it is easily comprehended.
The explanation presents some difficulties, though I have
seen them build enough to create an annelidan city; but
we will try to make it clear how the material for the construction of this little ocean tenement is hoisted up and
placed in exactly the right position to complete its circular
walls.

In the first place, the creature has some twenty or thirty
long, hair-like arms, which it propels out of the end of its
tube. Extending these in every direction and to an incredible length, they become so attenuated as to be scarcely dis-
cernible in the water; but these fine, delicate cords or filaments, hardly discoverable by the unassisted vision, may be
considered the ropes or tackling of its machinery for collecting the material which it needs for its sheath-like dwelling.
Suppose a grain of sand, for instance, is lying at some distance from the animal: by some sense it perceives it, determines to appropriate it, and immediately sends forth one of
its long, slender threads—*over* it or *to* it, for the extreme
points are so fine as to be distinguished with difficulty, but
the grain is reached. Watch it closely now! See! the bit
of sand begins to move gradually along and upward, *gliding
upon the surface* of this serviceable, rope-like filament. Observe, it is not grasped pincer-like with the end of the filament, but rides upward on the thread, like that mysterious
little wheel which thousands of our citizens see daily, creep-

ing up and over the wire which is one day to be a strand in the great cable of the East River Bridge. What the propelling or attracting force is, which causes the grain of sand to rise up against the laws of gravity and approach the mouth of this annelid, I have not yet been able to discover; but in all probability there is a system of muscular contractile organs in this fine filament, which a sufficiently strong magnifying-lens may yet bring to observation and recognition. Be that as it may, we will in the mean time watch for what we can *see* of this process, and we find that when the object has reached the end of the filament it is placed for a moment in the mouth, where it is evidently coated with a glutinous mucus and is then passed out again, and finally deposited upon the edge of its walls. The true level is kept, one side being built up at exactly the same rate as the other, so that no excrescences are left on the edge, but when finished all is of a uniform and even surface. The general appearance of the animal when at work forcibly reminds one of an immense *derrick*, full-rigged and in vigorous operation.

Nereis.—This is another member of the annelid family, but I cannot indorse it, either for good looks or excellence of behavior. Indeed, I hardly know of one redeeming quality to which it may lay claim. Possibly, however, it may possess some hidden virtue, which may yet redeem its character, and give it a higher place in my affections. Specimens of *Nereis* are found in nearly every group of serpulæ, generally hidden away among their closely-twisted tubes or in holes under stones, as if aware that it was not a favorite, and need not be on dress-parade. It appears to live upon the organic matter mixed in the mud and sand, for of these substances it consumes large quantities. But, though in the daytime limiting itself to this unsavory diet, at night it puts on another character. After dark it emerges from its concealment, swimming freely about, prowling and foraging among the delicate young fronds of algæ, and making sad havoc with its forceps-like jaws, cutting right and left like a pair of sharp, heavy shears, inexorable as those of Atropos.

Still another little house-builder is often found amid the tubes of the serpulæ, which indeed seem a sort of rallying-point, or chieftain's house, for the whole clan of the annelids. This latter variety is more simple in its organization, and appears to have much less machinery at its command; but it is equally as industrious as the best provided, though it has but two hair-like arms.

MOLLUSCA AND CRUSTACEA.

Among the mollusca there is one very pretty little shell-fish which is found quite frequently at Wood's Hole, Massachusetts, and which was brought to me by the well-known and enthusiastic collector, Mr. A. W. Roberts. This has afforded me many hours of entertainment from its curious and intelligent proceedings. It is the *Anachis similis*, described by Prof. Verrill in his valuable work on the mollusks of the New England coast, published by the United States Government in 1871–'72. The shell has ten whorls flattened; the colors are exceedingly variable, ranging from reddish-brown to chestnut, a light-yellowish brown, more or less mottled and speckled with white of a dullish hue; a band of white encircles the last whorl.

This apparently insignificant mollusk deserves more than a passing notice, and will well recompense the close observer of its habits; for, besides intelligence, it evidently has affections, which it proves by its love of society. In its internal structure there appears to be a small cordage-factory, for it carries about with it a life-saving apparatus, which it has it-self woven, in the shape of a fine silken cord. Should it be placed, or accidentally find itself, upon a piece of rock too high to slip off without injury, it unreels its silken cord and carefully lets itself down, instead of tumbling off at the risk of breaking its shell, as some of the more clumsy mollusks would do. When it wishes to ascend, it raises itself by the same apparatus, reefing in as it rises, like a sailor hauling in a rope.

It is also gregarious. Having had a number of specimens for years, I have scarcely ever observed one alone; and having many times separated them by placing them at opposite sides of the tank, I was certain in the course of a few hours to find them all together again, feeding or quiescent. A very peculiar trait with them is, that they are ready and willing at all times to be taken from the tank and handled; at least they show in no way any repugnance to this treatment. This is as rare as it is an agreeable feature; for marine animals are mostly very shy and sensitive to the touch of human hands. They appear also to be perfectly harmless, molesting none of their neighbors, simply enjoying their own life in a harmless and pleasant way, forming quite a contrast to the little prowling *Nereis*.

The *Anachis* lays its eggs in a mass, carefully covering them over with a clear, gelatinous secretion of a perfectly transparent and glass-like nature, which has the quality of thoroughly protecting them from the many voracious mouths that constantly surround them in their native habitat.

The *Lunatia heros* or *Natica heros*, as it is sometimes called, is another mollusk of very interesting habits. It may be found upon the sand-flats at low tide off Coney Island—indeed, upon almost any sandy shore from Maine to Florida, preferring those localities most thoroughly and freely exposed to the full force of the ocean-waves; their wildest fury has no terrors for it. Burrowing down a short distance in the sand, its power of suction is so great that the Atlantic billows pour over it in vain. If the *heros* goes, it is because the shifting sand goes with it, not because it is forced from it. It is not easily removed from its position by the hand, if its power of suction is applied to its fullest extent. It appears to find its food also in the sand, whatever its prey may be. When its soft, fleshy parts are fully expanded, it almost entirely covers its shell. Its foot or operculum is quite broad and large, and deeply concave, which of course gives it its great adhesive power.

This creature's manner of arranging for the protection of its eggs is exceedingly ingenious and interesting; no finished artisan could exceed the accuracy or quality of its work. The eggs are usually deposited in the pools left by the retreating tides upon the sandy flats, and a cursory observer would never suspect what they were or whence they came. They are laid in a broad, ribbon-like shape, arranged in a circle and mixed with sand, so that they look more like a tiny wash-bowl with the bottom out than anything else; and this combination of eggs and sand is finished off into a glassy smoothness. Should you hold this curious ribbon up to the light, you would see plainly and distinctly about a thousand eggs, for nearly every nest contains at least that number.

Our *butcher* mollusk is commonly known as "the drill;" its scientific name is *Urosalpinx cinerea*. It has a small, roughly-carved shell, and is often found upon the shores in the vicinity of New York, and also upon the coast of Massachusetts. They are not so numerous as some other varieties, but quite plentiful enough to make the localities they inhabit rather unhealthy for the oysters and other bivalves upon which they freely exercise their professional skill. It bores or drills a small neat hole through the hardest enameled shell, making an orifice as round and perfect as if executed by the modern diamond-drill. This preliminary accomplished, it sucks the unfortunate victim's substance away, leaving the empty shell upon the shore, with this professional death-mark upon it.

Purpura lapillus, famous for having anciently yielded a royal die, abounds on the shores of Massachusetts, but is not found much if at all south of that latitude, on this side of the Atlantic. Its most striking peculiarity is its mode of nidification. It lays its eggs in little goblet-shaped capsules, and seals the top of each with a perfectly-fitting cover, which the washing and beating of ocean-waves for several weeks fail to loosen or disturb. At the end of about three months this lid is at last unlocked, the cover lifted, and from each capsule tumble out five or six young baby shell-fish to take their chances in the great ocean of life; but, young and inexperienced as they are, they understand at once that to live they must eat, and may be seen immediately after they are freed clinging to the rocks and eating the sea-weed. Their soft, white, shelly covering is at first no larger than a pin's-head; but this shell very soon acquires a stony hardness, and reproduces the colors of its progenitors.

I would here call the reader's attention to the little door or *operculum* of some of these univalve shells, as it is not generally understood that this little calcareous object, which is attached to the animal, and serves to close up the orifice of the shell after it has retreated, is the *eye-stone* of commerce, such as the druggists sell, which is perhaps as precious to man as to the little animal; at least, if you were suffering with a miserable cinder or the like in your eye, you would promptly confess it. But for this wise and beautiful provision of a door of stony hardness, the little helpless mollusk would be destroyed by the thousands of hungry mouths that surround him in his ocean-home. This operculum or " eye-stone " is convex on one side and flat on the other, and, when drawn in by the animal, shuts the opening of its shell like a valve, perfectly air and water tight, varying in size from a pin's-head to *three inches* in diameter. (We do not, however, use the largest size for eye-stones.) The colors of some are very rich, and these have been used quite extensively for articles of jewelry, such as sleeve-buttons; mounted in fine gold, they make a very pretty article. If you place one of these stones in a saucer containing some weak acid, it will move around like a living animal. This is caused by the evolution of carbonic-acid gas, contained in the carbonate of lime of which the stone is composed. Loaves of bread are said to move sometimes in the oven from the same cause. Undoubtedly the eye-stone will often remove dust or any foreign substance; if introduced under the lid of the eye, the stone is shifted about by the motion of the muscles, and any little particle it touches will adhere to it, and be brought out with it.

The *Pecten irradians*, the "St. James's shell," best known as the common *Scallop*, is almost as familiar a sight in our markets as the oyster; but, though chiefly regarded for its edible qualities, its interesting habits form its principal attraction to the zoölogical connoisseur. Its history is unique—its fame wide as Christendom. Being very plentiful on the shores of Palestine, it became customary for all European pilgrims, on their return, to attach a scallop-shell to their dress to prove that they had really been to the Holy Land. So thoroughly was it identified with Christian knighthood and saintship, that it became the insignia not only of the great apostle whose characteristics the monks changed from fisherman to warrior, but the recognized badge of several of the half-saintly, half-chivalrous orders of knights which arose and flourished in the middle ages.

But, not to dwell on its illustrious associations, or even on the intrinsic beauty of the shell itself, we will bestow our attention upon the inhabitant, which, if it

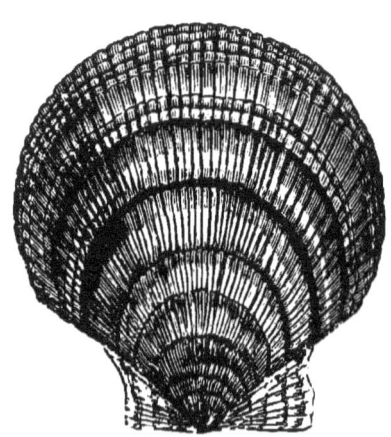

SCALLOP (*Pecten irradians*).

prided itself on its ancestry, might certainly boast, over all the mollusks, of its *sang azuré*. All shell-fish have over their bodies and beneath their shells a flexible tissue known as their "mantle;" it is, indeed, from the secretions of this mantle, of many folds, that the shelly carapace is originally formed. In the case of the scallop a portion of this mantle can usually be seen, showing a finely-fringed curtain of scarlet or orange, the mantle itself being of a delicate fawn-color, the whole set off with a number of bright, glistening eyes, of an elegant emerald-green, encircled with a band of turquoise-blue. The finest jewels of our fairest belles can be no brighter than the natural adornments of this common mollusk. In their native element alone the scallops show to perfection all the beauties Nature has lavished upon them, especially when seen in motion. They move in a rapid zig-zag fashion, and with the speed of an arrow, the propelling force being secured by the rapid opening and shutting of their valves. One can scarcely see a lovelier sight than that of a large number of these pretty creatures, with shells of every hue, from purest white to black, enlivened with shades of pink, yellow, fawn, and other tints, darting about in the clear water, up, down, here, there, everywhere. In their flight-like movements, vertical, horizontal, east, west, north, and

south, they are more suggestive of a flock of winged animals than of bivalves of which to make a meal. When at last they dispose themselves to rest, sinking to the bottom

THE DANCING SCALLOPS.

for that purpose and there remaining passive for hours at a time, they will in the aquarium, if not properly managed, come to anchor by tying themselves with their byssus to the rocks; and, if that occurs, they will entertain us no more with their lively and amusing habits.

DECORATING CRAB, OR SEA-SPIDER.

The Sea-Spider, or Decorating Crab (Libinia canalicu-
lata).—Of all the crab tribe, this is surely the most fantastic
little fellow, and ought to be considered the "missing link"
which unites the animal creation to the human, for he has
certainly one of the first instincts of civilization, namely,
that of attempting to cover himself with extraneous and
ornamental garments. He is the dandy of the sea. Bits of
sea-weed are his great reliance, but small objects of almost
any kind he will appropriate, even to pieces of stone or wood.
One of mine showed considerable taste and an idea of style,
preferring always the most gaudy colors which he could find
in the tank. These animals will spend hours every day at
their toilet, appropriating with their hand-like claws bits of
sea-weed, *Sertularia*, sponge, or *Tubularia*. One will perhaps place a bit on the tip of his nose, or suspend from it a
long, ribbon-like strip of red or green algæ, or affix similar

fragments to his legs,
elbows, or knees, as
we may call them.
He does not appear
to take these pieces at
random, but has the
air of selecting them
with care, and then
leisurely cutting them
off from the large
fronds with his own
nippers, of which he
has two pairs, one
upon each of his two
foremost arms. Hav-
ing severed the de-

THE DECORATOR.

sired portion, he takes it up in one of his hands (for his nippers serve for hands as well as shears), and, placing one end
of it to his mouth, evidently deposits upon it a species of

mucus or marine cement, which secures the object in the position in which his lordship sees fit to arrange it, and in which matter he is somewhat fastidious. This mucus must have great strength, for in his native element he will walk about thus arrayed, without any danger of his ornaments being washed away even by the rolling surf. In the tank, when his toilet is completed, he will advance to the front or most conspicuous spot he can find, and as near to the spectator as he can conveniently get, with a self-satisfied air, as much as to say: "I'm in full dress now; how do you like my style?" I have also had some of these "decorators" who showed a sort of paternal affection for the young actiniæ, each seeming to take a particular pride in placing them, sometimes five or six at once, upon his back, among the bits of algæ already there, and then parading round as if bound to give his pets a free ride.

SHEDDING, OR EMPTY SHELL OF SPIDER-CRAB.

At certain periods, like all crustaceans, this spider-crab becomes too large for its shelly covering, and is obliged to move out, or rather its house is moved away from it. This is effected by a rupture of the tissue connecting the upper and lower carapace, near the hinder part. The body is first slipped out through this opening, then Mr. Crab slowly draws out his arms, one after the other, just as a man would withdraw his from his coat, and so proceeds until the ten arms are all freed. Its old habitation is left perhaps on a rock,

or hanging to a sea-weed, and the most practised eye might
be deceived and take it for a living crab. So perfect and un-
injured is the abandoned skeleton, that even the delicate coverings of the eyes and antennæ are all there, standing erect
and staring at you like a ghost. Although I have witnessed
this interesting operation probably hundreds of times, I was
completely deceived by a specimen in my tank only a week
ago. Mrs. Damon calling my attention to a new sea-spider,
I saw him hanging in a frond of *Solaria*, looking boldly out
of the tank at us; and I supposed for some time that some
kind friend had contributed another of these crabs to my
already overstocked tanks. But soon the mystery was ex-

CRAB EATING A CLAM.

plained, for just below I espied the familiar face of my old
sea-spider in his new and undoubtedly more comfortable
suit, for now he was at least one-third larger than before the
shedding. He is very shy now, and most unobtrusive and
polite; for his new shell is yet very soft and tender, and he
has all he can attend to in preventing the other animals from
eating him up, house and all. Just after the moulting, these
creatures usually half bury themselves under some rock or
weed, until the new shell is hard enough to inspire them
with courage and confidence.

Our common lobster, *Homarus Americanus*, has the

same habit.   So also have the common blue or edible crabs,
which, in the transition-stage, are sold in the market as "soft-
shell crabs," and are by many supposed to be a distinct species.   The fishermen recognize the fact of their identity by

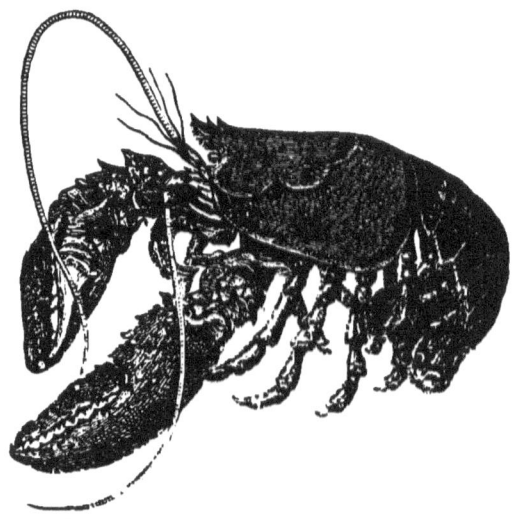

AMERICAN LOBSTER (*Homarus Americanus*).

the name they give them, "shedders."   I cannot say positively how often this moulting takes place, but during the
earlier portion of the crab's existence it is probably not less
than four times a year.

For scavengers we have quite a variety both among the
crustacea and the mollusca.   The crabs perform the coarser
or more arduous duties, and the buccina give the finishing
touches, sweeping up the finer fragments that would otherwise be left to decay.   These animals are almost indispensable
to the healthful condition of sea-water in the aquarium; at
least they are invaluable in preserving its purity.   They are
mostly carnivorous, living almost wholly upon dead animal
matter.   (I have known them to "mow" off confervæ from
the sides of tanks; but, fortunately, my tanks are not troub-

led with that annoying growth.) Cast a dead fish or crab in
their path, and it will be entirely disposed of in a few hours.
We are here tempted to ask if an army of buccina would
not be a valuable acquisition for our Street Commissioners;
but think what the poor things would have to eat! Of all

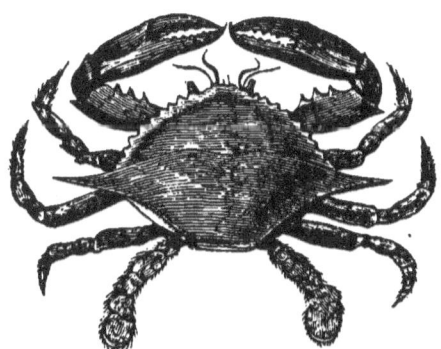

EDIBLE CRAB (*Callinectes hastatus*).

the shell-fish our coast supplies, I believe this is furnished in
greatest abundance. One can hardly walk a yard upon any
part of our sea-shore without treading upon them; every
creek in the adjoining meadows as well swarms with them.

COMMON WHELK OF GREAT BRITAIN (*Buccinum undatum*).

I have seen them at Unionville, Long Island, at low tide, by
the million, or in heaps by the bushel, if you prefer them by
measure. They deposit their eggs mostly upon the plants
near the sea-shore, in little capsules, semi-transparent, and
securely fixed to whatever object they adhere to.

As a representative of the learned professions among our
marine curiosities, we have a surgeon, commonly called the
"doctor-fish" (*Acanthurus*), from the fact of its having a
lancet-shaped weapon situated in a longitudinal groove upon
each side near its tail; and it has both the skill and power
to use it most effectually when occasion requires.   A more
complete description of this remarkable fish will be given in
the next chapter.

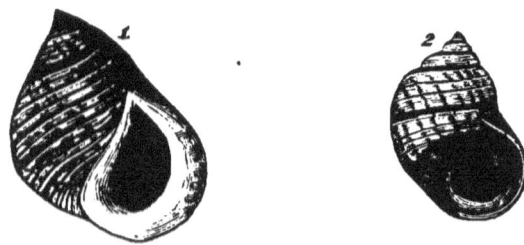

PERIWINKLE.—1. *Littorina littorea;* 2. *L. rudis.*

CHAPTER V.

WE need not leave the shores of the United States to find either beautiful fishes, crustacéa, or other marine curiosities. Yet here on this beautiful little island,[1] situated in the middle of the Atlantic, eight hundred miles from any other land, it is natural to presume that we shall see novelties not easily attained in our own country. Nor are we disappointed. The very first morning after our arrival, between four and five o'clock, we become conscious of some slight local excitement, and find on making observation from the veranda of our hotel, the Royal Victoria, that this slight ripple upon the usual repose of semi-tropical life is caused by the arrival of the fishing-boats, which have just come in, and toward which the early customers are proceeding, to make their purchases.

These people, mostly colored, are of themselves a sufficient novelty to attract the attention of a stranger, as they come along chattering and singing with that nonchalance, and evident absence of care, so different from the anxious and busy look which the American almost always bears about with him. *They* will not easily pass from our remembrance.

But there is doubtless something in those boats which will interest me more than even these odd specimens of humanity; so I hurry along with the gathering citizens of vari-

---

[1] These observations were made during a residence at Hamilton Harbor, the largest port of the Bermuda Isles.

4

ous shades, from Caucasian to African, and proceed to in-
spect what is being offered for sale. For these fishermen are
also the retail dealers, and sell their cargoes as quickly as
they can directly from the boats. Here there is no anxiety
as to "whether the fish are fresh," for, besides their being
so recently caught, each boat is provided with a deep well or
permanent tank into which "the catch" is thrown and there
kept alive until sold. On this occasion there were about
twenty of these boats. Several of them were loaded with
brilliantly-colored beauties, which it seemed a shame to sell
by the pound, to be *eaten*.

One of the most common varieties for market-use is the
large-mouthed grouper, *Serranus striatus*, marked with white
and black spots, changing to a brown or pale white with almost incredible rapidity while one is gazing upon it.

In a letter which I received from the late much-loved
and highly-esteemed Prof. Louis Agassiz, in relation to the
fishes of Bermuda, some of which I am about to describe, he
says :

"This collection is highly interesting in a scientific point of
view, as it shows the ichthyology of Bermuda to agree completely with that of the Gulf of Mexico, and not at all with
that of the Middle States or of Europe."

The angel-fish, *Holœcanthus ciliaris*, is another variety
largely used for the table; and this, too, appeared like a
desecration in my eyes. When alive their bright colors vie
with, if they do not exceed, the most brilliant plumage of
the Brazilian humming-bird; the rainbow's tints are pale in
comparison. They seem to me to have more expression in
their countenances than any other fish. I always fancy as
they swim toward me that there is almost a *smile;* there
certainly is an amiable air, corresponding with their name.
They have a somewhat lazy, slow, and careful style of swimming, with their long, ciliated fins trailing after them, as they

glide in and out among the coral-branches with perfect ease and grace, often reminding one of· some of the graceful feathered tribe—if we could imagine the latter swimming without draggling their plumage.

The angel-fish is of a broad oval shape, and remarkable for its crescented caudal extremity ; its forehead is very high, but its human look as it approaches one is to me its greatest attraction.   The scales are very strongly defined, and are laid in with mathematical precision, reaching high up on the dorsal fin, their size gradually decreasing until individuality is lost in the extreme fineness to which they are reduced.   A band of deep velvety blue adorns the high forehead, extending backward and along the dorsal fin, gradually emerging into a rich golden-yellow tint on the caudal.   But what an exquisite blue !   Did any one ever see just such a shade ? It is unequaled in silk or satin, and many a young lady would be happy indeed could our beautiful "angel" impart its secret, and tell us how to imitate its unrivaled blue.   One of these beauties I have seen with a crescent-shaped black marking across its body, and with the same rich blue coloring on the tips of its operculum or gill-covers.

To see these lovely creatures laid out for sale on the market-stands seems almost wicked.   But the Bermudians have no conscientious or scientific scruples, and fry them with as little compunction as a New-Yorker would an eel.   I have brought many of them alive to the United States, swimming in pure sea-water, and have by most constant and careful attention succeeded in keeping some of them for nearly a year. In the winter, when the temperature was low, I found it necessary to protect "my angels" from the cold ; and not venturing to tamper with the water, I adopted the expedient of wrapping the tank with woolen blankets, thus keeping in the natural heat, and graduating the temperature as nearly as possible to that of the Bermudian waters.   These fish, though too handsome, in our opinion, to be eaten, are not

themselves too good to eat others; their preferred morsels
are mollusks and crustacea.

Sarothrodus bimaculatus—called by the various names
of " the bride," " peacock-fish," and " four-eyed fish "—is
another perfect little gem ; and at the first glance one is forcibly reminded of some bright piece of jewelry.   In form it
is nearly oval, flattish, with its mouth elongated into a sort
of tube-like projection.   It has the reputation (like the chæ-
todons of tropical waters, the chelmons of the Asiatic seas,

CHÆTODON ROSTRATUS.

and the archer-fish of Japan) of shooting flies or other insects
which it may perceive on the rocks, or on the plants near the
shore, above the surface of the water, by forcibly ejecting a
well-aimed drop of water at the intended victim.   This interesting performance I have never had the pleasure of witnessing, except as illustrated, on *paper*.   The coloring and
marking of this fish are also very ornamental.   Its body is a
grayish pearl, with a band of jet-black crossing the forehead,
and running directly through the eyes ; another similar band

a quarter of an inch in width crosses the caudal fin in a circular direction. Did the description stop here, we might say that it was dressed in "half-mourning ;" but an ornamental design which it bears upon the hinder portion of the body precludes the "mournful" theory. These peculiar markings consist of two brilliant round spots, about the same size as, and bearing a perfect resemblance to, the eyes of a peacock's tail-feather. From these and its proper true eyes this little fish has acquired the name of the four-eyed. Its style of swimming is in direct contrast with that of the angel-fish, for it dashes about swiftly and nervously through the water.

Holocentum longipenne, or the "squirrel-fish," probably obtained its name from its habits, which are as closely imitative of the squirrel as those of a marine animal can well be. He is not seen swimming about as a fish ought to do, but hides himself in holes or crevices of the coral-banks. Looking over the edge of the boat, as we sail over the abounding coral-groups, one may almost always observe at some point, often at many, two big black eyes staring out of the nooks and crannies in the rocks. If it is your first experience you naturally look again, and wonder what the eyes belong to, for that is all you see. You conclude, of course, that it is some sort of fish, and the next thought is to draw this "hide-and-seek" fellow out of his hole. So you bait a hook with some tempting morsel, and throw out your line. How he rushes! with lightning-speed he has darted out of his hole and seized the bait. Had you time to think, you would be sure you had him ; but quicker if possible than he dashed out does he drive back again, carrying hook and line with him, twisting and fouling the latter around and between the coral-branches, or some projecting peak, so that you may as well say "good-by" to hook, line, and squirrel together. The color of this brisk little fellow is of a bright, shiny red, and it has large, beautifully-serrated scales. Its dorsal fin is immense in proportion to the size of the body. Its gill-covers are armed with long, sharp, dangerous spines.

On account of this habit of the squirrel-fish, of making for his hole the instant that he feels the hook, and the difficulty of managing a line among the projecting coral-branches, the natives never attempt to take this or other fish in that way. All they take are caught in *traps*, very simply con-structed, but quite effective. They construct a sort of crate or basket out of. strips of wood, of irregular and ingenious shape, but usually broader than high, and holding from one to two bushels. Having put in their bait—meat, stale fish, crabs, lobsters, mussels, almost anything—they securely attach to it a long and strong rope, and proceed with it in a boat to some good fishing-ground, generally near some reef at a distance from the land. Dropping the trap overboard, they let it sink to the bottom, not forgetting to secure the free end of the rope to a buoy to indicate its position. Usually several of these traps are deposited at each trip, at different points. They are allowed to remain for about twenty hours, when the fishermen visit them, haul the traps aboard, cursorily examine the contents, and, if nothing very dangerous is seen, open a little door in the trap and turn out all these mixed-up victims, which have been unfortunate enough to find themselves hopelessly ensnared. The bait is then renewed, and overboard goes the basket again, ready for another day's work.

It is always interesting, sometimes amusing, to witness the operation of emptying the traps, for occasionally they catch a very ugly customer. An octopus or "devil-fish" is sometimes found secreted in one of these traps—a very dis-agreeable and, when of large size, dangerous animal to deal with. For the present we will throw him overboard again, but will take an opportunity hereafter to describe and illus-trate him more particularly.

But there is one snake-like creature, the green maray,

with his relative the spotted maray, which creates much more consternation among the occupants of the boat than even the discovery of a "devil-fish" among the contents of the trap. These are not infrequently found in the Bermudian waters; they have a wicked, snaky eye, and are very dangerous. Should one of these creatures come tumbling out of the trap, out would go the negroes over the other side of the boat into the sea! Indeed, all the natives have a wholesome fear of this creature, and put themselves at a respectful distance from it with what speed they may. These fish vary in length from three to ten feet, are armed with crooked teeth like the anaconda, and have a pointed nose and a large, fleshy dorsal fin running nearly their whole length. Their predominant color is a rich olive-green. When an animal which inspires dread is really ten feet in length, imagination readily adds to its true measurement. May not this maray be the original foundation for the "veritable sea-serpent," which varies so continually in size and appearance? Or are these the degenerate and diminutive descendants of that rarely-seen and *never-captured* tyrant of the ocean?

Hon. C. M. Allen, United States consul here, has a fish-pond and fountain, which is supplied with pure sea-water by means of the tide flowing in and out; and by a very ingenious mechanical arrangement, invented by himself, the water is thrown some thirty feet high into the air. This fountain will act about twenty-three hours out of the twenty-four. Mr. Allen has just informed me that he has in his pond some very fine specimens of the maray, and that they exhibit an incredible amount of strength in their teeth, by biting and pulling out the stones from the walls of the pond.

The waters of Bermuda are so prolific in strange and wonderful fishes, that we with difficulty make our selection between the many we must omit to mention and the few we have space to describe. There is one which should, I think, be called the "knight in armor," but which has had fastened upon it the utterly prosaic name of the "cow-fish"—*Ostracion sexcornutus*. This fish is covered with a perfect coat of shelly armor, excepting only two round holes for its eyes, and sufficient apertures for its fins to pass through; and thus rigged, it is invulnerable with the exception of its steering

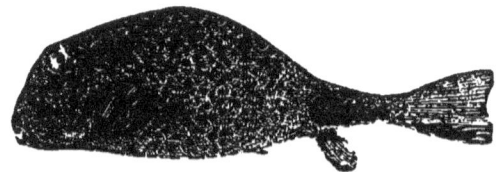

YALE'S TRUNK-FISH (*Ostracion Yalei*).

apparatus and its eyes. In shape it is also a curiosity, being formed like a three-sided trunk (hence also called trunk-fish), and when swimming with its flat side downward, its back somewhat resembles the sloping roof of a tiled house. It has two horn-like projections on its head, which suggest some resemblance to those of a cow; hence its name.

In addition to its singular armature this fish has an extraordinary power of change as to color, changing from black to blue, or to white and green, in a few minutes. Indeed, the transformation is so rapid that it appears magical, and one feels almost compelled to believe that he is looking at different specimens, which have been mysteriously substituted for the original. The intelligence of the creature appears to be very limited, much inferior to that of most fishes; indeed, I think it may even be called stupid without doing it any great injustice. At least, it has not sense enough to get out of the way of an ordinary row-boat in open sea. It may be, however, that its armor-like covering is such a protection from ordinary dangers as to give it a degree of confidence which less well-provided fishes cannot feel; and with boats and men they are not supposed to be sufficiently acquainted to appreciate the danger which threatens all the lower orders wherever these appear on the scene.

To illustrate this lack of intelligence, I will relate my experience in capturing my first prize. But it may be well here first to explain that the extraordinary clearness and purity of the waters among these coral islands is such that objects may be clearly discerned at a depth of twenty fathoms —a fact which adds intensely to the pleasure of sailing upon them; for, whether a naturalist or not, one cannot fail to be charmed at the variegated views of coral caves, branching tree-like forms, richly-colored algæ, shells, and the vast variety and beauty of the marine creatures, fishes and others, which are to be seen gliding about in these submarine bowers. As I was one day seated on the bow of our boat, I saw a very fine specimen ahead, which naturally excited the attempt to capture it: and this proved a very easy task; for I actually *drove it along* by means of an oar toward the sandy shore, until it became stranded in the shallow water, where it was immediately secured with my hands, alive and in perfect condition. Possibly this docile disposition may have served to aid in fixing its common name upon it.

The parrot-fish is another of those noted for its " good looks." Its scientific name is *Scarus turchesius.* Vaulting from the farm-yard to the Brazilian groves in search of a name, our naturalists have not greatly erred in selecting that of parrot for this animal, for there is certainly a most striking resemblance in the head of this fish to both beak and head of our common parrot; its colors, too, measurably correspond to those of one of the best-known species of parrots, it being of a bluish-green tinge. This fish does not grow to any considerable size, nor is it esteemed for the table. Neither does it thrive in confinement. I have never succeeded in keeping one over three months, and can now only revive my memory of its beauties by regarding the skeleton form of one which I have preserved among many other Bermudian curiosities.

Cancer cranium morta, or the "grave-digger crab," is one of those singular creatures whose home may be said to be the sea, but which yet spend much of their time on the land. Its habits differ from those of many of its crustacean relations, particularly in its nocturnal perambulations. In the daytime it lies concealed in holes which it digs in the earth, formed something like a wild rabbit's burrow. It is useless to seek this species in the daytime; the only opportunities for capturing them occur at night, and the manner of doing so is rather a novel method of warfare.

Suppose our party formed and ready for the expedition, at about twelve o'clock on a bright moonlight night. We start from Hamilton Harbor in a small boat, and row across the bay-like indentation of the shore a distance of about three miles, where we land, haul our boat high up on the beach, and proceed on foot over a neck of land a mile or so, until we reach the "North Shore." Here, on the flat lands, we shall find our "hunting-ground." Now the order is given to "halt!" and each one is requested to keep perfectly still, as that is our only chance for catching sight of our game; for any unusual noise would quickly alarm them, and cause them to retreat to their burrows. A few moments of hushed silence. "Listen!" Here they surely are, marching about as bold as lions, yet cowardly at heart as hares—unless cornered, with no chance of escape, when they will make a desperate show of fight, and will even advance upon their pursuers with their large fore-claws uplifted in very valorous fashion. What do you suppose is our implement of attack? You will never guess. It is a *blanket!* One of us takes this, and places himself before the hole which the crab inhabits. When, alarmed by the others of the party, the crab makes for his retreat, we deftly throw the blanket over him, rolling him up, and taking good care that he does not get a claw free and a chance to grip our hands or arms. When quite secure, we throw him into a large bag prepared for the purpose, thus literally " bagging " our strange game. Having in this way seized all we care to capture, we return to our boat, and turn the bow homeward, after one of the most odd and exciting adventures which this island affords.

These nocturnal crabs have gained for themselves, among the natives at least, the reputation of *digging into graves;* but I scarcely believe so badly of them as that legend would imply. One reason for my disbelief is that they would have to dig too deep, and another that there is no necessity; they can find food enough without taking so much trouble. However, in their foraging expeditions at night they do make away with vast quantities of dead animal matter, fish, or anything else. In this they are probably performing a beneficial work, ridding the land of impurities which the hot sun would otherwise unpleasantly develop.

CHAPTER VI.

*THE OCTOPUS, OR DEVIL–FISH, AND ITS CONGENERS:
THEIR HABITS, TRUE CHARACTER, AND MODE OF
CAPTURE.*

THE transition from angel-fish to devil-fish is no greater in nomenclature than in fact. The creatures themselves are as different in appearance and habits as the contrasting names imply; for, without attributing moral qualities to animals who simply fulfill the functions which their special organizations demand, it is yet impossible not to feel an attraction toward the pleasing and beautiful, and a certain degree of repulsion toward the ungainly, secretive, and diabolical-looking octopus.

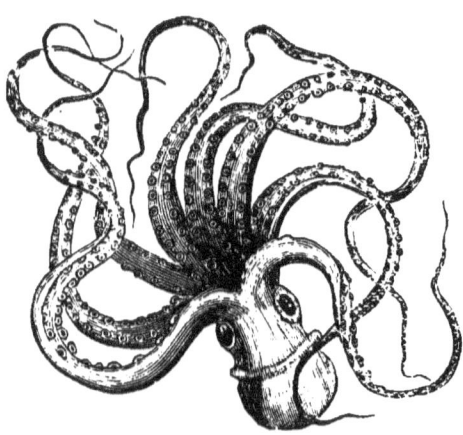

OCTOPUS OR DEVIL-FISH (*Octopus vulgaris*).

Perhaps no better introduction to this chapter can be given than to recall to the minds of our readers the terribly vivid description of the devil-fish by that grand master of romance, Victor Hugo; for, though incorrect in several scientific details, the general description is the best we have had, though Jules Verne's is almost as dramatic and nearer to Nature. In "Les Travailleurs de la Mer" M. Hugo says:

"To believe in the existence of the devil-fish, one must have seen it. Compared to it the ancient hydras were insignificant. Orpheus, Homer, and Hesiod, *imagined* only the chimæra — Providence *created* the octopus. If terror was the object of its creation, it is perfection. The devil-fish has no muscular organization, no menacing cry, no breastplate, no horn, no dart, no tail with which to hold or bruise, no cutting fins, or wings with claws, no prickles, no sword, no electric discharge, no venom, no talons, no beak (?), no teeth. It has no bones, no blood, no flesh. It is soft and flabby, . . . a skin with nothing inside of it. Its under surface is yellowish; its upper earthy. Its dusty hue can neither be imitated nor explained; it might be called a beast made of ashes which inhabits the water. Irritated, it becomes violet. It is a spider in form, a chameleon in coloration.

"Seized by this animal," he adds, "you enter into the beast; the hydra incorporates itself with the man; the man is amalgamated with the hydra. You become one. The tiger can only devour you; the devil-fish *inhales* you. He draws you *to* him, *into* him; and, bound and helpless, you feel yourself slowly emptied into this frightful sac, which is a monster. To be eaten alive is more than terrible; but to be *drunk* alive is inexpressible!"

This overwrought but wonderfully dramatic description (but a small part of which we have quoted) at once excited a popular interest in the habits and history of the octopus, though it was well known and described by Aristotle before the Christian era. Moreover, the animal so graphically pictured by the novelist was a mere "baby devil" in comparison with many which exist, and which have been described by that enthusiastic naturalist, Prof. Verrill, of Yale College.

In a letter addressed to me on this subject by Prof. Spen-

THE GIANT SQUID.

cer F. Baird, under date of April 1, 1878, this distinguished naturalist says:

"The giant squid in the New York Aquarium can only be designated as an infant or dwarf in comparison with the gigantic species of the Pacific Ocean—those upon which the sperm-whale is known to feed. Chunks of squid-remains are not infrequently found in the throat or stomach of the sperm-whale, apparently indicating specimens from ten to fifty times the size of the Newfoundland variety. I was informed that a considerably larger specimen than that at New York was cast ashore at Newfoundland later in the season. The arms of the latter, if I recollect right, were some ten feet longer than those of the other."

The specimen referred to by Prof. Baird as at the public aquarium in New York, is of the species known as *Architeuthis princeps*. It measures about forty feet, and is preserved in alcohol. I have in a bottle some specimen portions of the sucking-disks, showing the serrated edges, from the arms of this terrible animal; and I have also a perfect specimen of a smaller species of the animal itself in my private collection.

Prof. Verrill's reports apply to the devil-fish found in our northern seas, and Prof. Baird mentions those cast ashore at Newfoundland; but that they are not limited to the northern waters is certain. The late Captain Frederick Reimer, of New Jersey, a very intelligent observer, who was in Beaufort Harbor, North Carolina, in 1862, described one that he saw there which measured fully thirty feet in length. Any one who has seen the specimen captured at Newfoundland can readily conceive how such a monster could stretch out its two long arms and seize its prey. These arms together form a pair of powerful pincers at their extreme ends, and are furnished for their whole length with two rows of perfect sucking-disks, or some two thousand air-pumps; the edges are also cut into sharp, saw-like teeth, as hard as steel, and these are buried in the flesh of its prey. With all these appliances it could easily reach a distance of twenty-five feet, and bring the body of a man to its mouth, where, with its powerful iron-like beak, it crushes the helpless form, and swallows or *drinks* it down, as Victor Hugo says.

My own experience with them has been principally in the Bermudas. They are there caught in basket-traps, formed of wood, described on page 60. With a trap baited with mussel, crab, or lobster, of which the octopus is particularly fond, we row along the island-shore, among the more rocky parts, until we discover some indication of his majesty's retreat. Their hiding-places can only be diagnosed by experts, but one of the trails by which they are traced is the presence of dead shells in unusual quantities, particularly skeletons of crabs, which will be pretty certainly seen near the water's edge, or at the mouth of the cave inhabited by a " devil." The clearness of the waters greatly aids in the search. When a promising location is reached, we throw overboard the trap, which sinks to the bottom of some ledge, or rests upon a reef of coral. A rope, which is attached to it, is secured to a buoy to mark its place on the surface of the sea, and it is left for twenty-four hours. Then we return and haul it up, and, if the place of deposit has been well chosen, we shall soon see the long arms of Mr. Devil protruding through the basket, searching and stretching in all directions, seeking to understand how it is that positions have become so reversed —that he is the captured instead of the capturing party. His color changes with anger and vexation, and his body then displays numerous bunches or tubercles, which always appear when the animal anticipates danger.

The trap being opened, we seize him quickly by what we must call the neck, the portion between the head and trunk, while his eight arms or legs, as you may choose to call them, are struggling and twisting in all directions, sometimes becoming attached to our own arms and twining about them. Those which I caught and handled personally had arms of remarkable softness and suppleness, so that their contact felt more like a running liquid upon my flesh than a structural substance; [1] and, indeed, though so formidable under certain circumstances, the preponderance of fluidity in their composition may be judged from the fact that I myself saw one, which measured three feet in length by five or six inches in width, squeeze or *run* itself through a crevice not over *half an inch* in width!

I should have mentioned that if it is desired to preserve the octopus alive, the pressure on the neck should not be too severe, for that is their vulnerable point; and a person attacked by one should never lose time in striving to loosen its arms, but grasp if possible this portion connecting the head and body, in which way they may be easily killed.

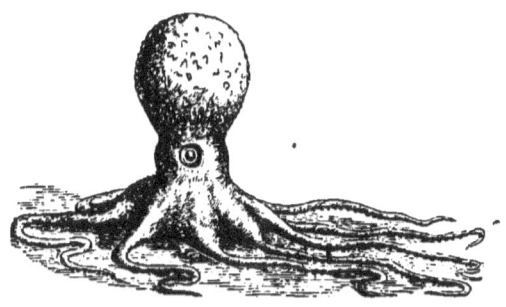

AN OCTOPUS RUNNING.

In regard to their powers of locomotion upon land, on which there has been considerable controversy, I can assure the reader that I have seen a full-grown octopus at the Bermudas spring up out of the water, only a few feet forward of the boat I was in, and run up a perpendicular rocky cliff for more than *two hundred feet!* This ledge of rock bore a general resemblance to our Hudson River Palisades at their steepest portion. We soon learned the cause of this seemingly strange performance, when we discovered one of those beautiful bright-red crabs, which are native to the locality,

[1] This lack of tension probably resulted from my pressure upon the neck.

trying to escape from the clutches of this devil-fish. The crab, being frightened almost out of its simple wits, had run up the rocks for safety; but its tactics proved sure death in the end. As to the speed of the octopus, it appeared to me to travel much faster than I could run. At least, I should not care, if unarmed, to engage in a race with one, unless Mr. Devil started a good way *ahead*.

In this case I soon came into closer acquaintance with our agile friend, for the next morning I had the satisfaction of discovering that his lordship had walked into our trap, which we had carefully placed near his cave; and now that we could see him face to face, we found that his strength was enormous as compared with his moderate size. Being placed in a bucket of water, such as is usually found on a ship's deck, he attached his eight arms to the bottom and sides, by means of its powerful and perfect-working suction-disks, so firmly that I several times lifted the bucket, water

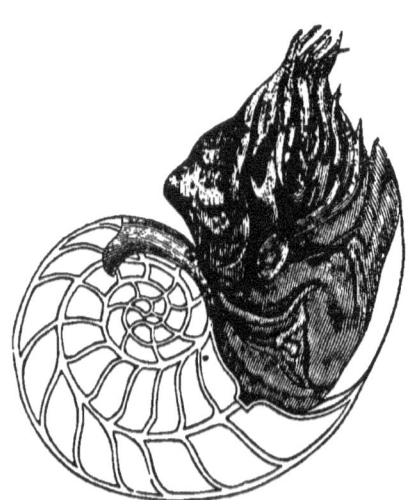

PEARLY NAUTILUS (*Nautilus pompilius*),
showing chambers inside of shell.

and all, by taking hold of the animal's body, and twirled it over my head. The more I twirled the more firmly it stuck. An octopus will not relax its hold " on compulsion," any more than Falstaff would " give reasons." It is as self-willed as some human animals.

According to scientific classification, the octopus belongs to the division of soft-bodied *Mollusca*, and the class of *Cephalopoda*— meaning "feet proceeding from the head." Of these the octopus, as its name indicates, has eight feet, or arms; for, though these long appendages are sometimes used as feet, they are habitually used as arms.

Of the octopoda family is the small paper nautilus or argonaut. How few of our readers who have admired this beautiful shell, with its mother-of-pearl lining, have realized that its former inhabitant was own cousin to the horrible devil-fish—a female cousin, we must add, for the shell is not connected with the animal organically, but is held in position by two of the long arms, with the sole purpose of protecting the eggs. The male argonaut has no shell.

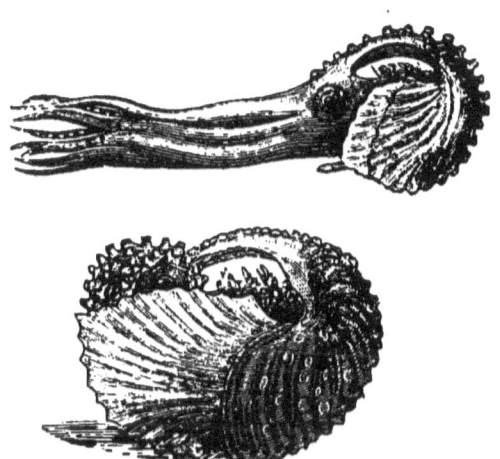

PAPER NAUTILUS (*Argonauta argo*).

Though all the octopods, large or small, can swim freely at will, such is not their habit; they prefer to lie concealed, or partially so, on the side or in the clefts of rocks. There the octopod's body is protected from the attacks of other animals, while it can extend its long feelers in search of prey, of which fish, mollusks, and crustaceans, are the principal objects. Its movements, when an object of food is perceived, are marvelously rapid, swifter than the flight of an arrow from the bow of an experienced hunter. The long, flexible arms grasp the victim; its hundreds of suckers, act-

ing like pneumatic holders, make escape impossible; and, as
the long arms draw the object nearer and nearer, the other
shorter arms add their multiplied disks, forming "a perfect
mitrailleuse of inverted air-guns, which take horrid hold,

ARGONAUT WITHOUT THE SHELL.

and the pressure of air is so great that nothing but closing the throttle-valve can produce relaxation." This throttle-valve is the neck, as we have before described. Those
lengthy appendages, the limbs, are rather in the way when the
animal is swimming, and would act as drag-anchors if left

ARGONAUT WITH THE SHELL.

pendent; but the octopus usually draws them close alongside,
whence they extend in a horizontal position, acting the part
of a tail to a kite. It propels itself by drawing in and ex-
pelling water through its locomotory tube. The octopus

swims backward, and it has been remarked that it changes its color to a darker hue when it starts out for a swim.

This changing of its hue, apparently at will, is one of the most peculiar characteristics of the octopus. It may be considered the chameleon of the sea. Its ordinary color when in repose is a mottled brown; but if irritated it assumes a reddish hue, approaching to purple. Nature seems to have been almost superfluously careful in furnishing this animal with protecting elements; for this coloring-matter, which resides between the inner and outer skin, enables it even to assume the color of the ground or rocks over which it travels, so that one can hardly say what color it is before it may have changed to something quite different. When exhausted after a battle or a struggle to get out of a trap, it turns pale, like a human being.

Some persons besides Victor Hugo's hero have had a chance to test the strength of these devil-fishes. Major Newsome, R. E., when stationed on the east coast of Africa in 1856–'57, undertook to bathe in a pool of water left by the retiring waves. He says:

"As I swam from one end to the other, I was horrified at feeling something around my ankle, and made for the side as speedily as I could. I thought at first it was only sea-weed; but as I landed and trod with my foot on the rock, my disgust was heightened at feeling a fleshy and slippery substance under me. I was, I confess, alarmed; and so apparently was the beast on which I trod, for he detached himself and made for the water. Some fellow-bathers came to my assistance, and he was eventually landed. . . . As the grasp of an ordinary-sized octopus holding to a rock is not less than thirty pounds, while the floating power of a man is between five and six pounds, I believe if I had not kept in mid-channel it would have been a life-and-death struggle between myself and the beast on my ankle. In the open water I was the best man; but near the bottom or sides, which he could have reached with his arms, but which I could not have reached with mine, he would certainly have drowned me."

The Major was right; he had every chance of sharing the fate of the immortal Clubin.

When a crustacean casts a limb from its junction with the body, it is after a time reproduced; if injured below this point, it has no recuperative power. But our " devil-fish," which really seems favored beyond its deserts, will reproduce any injured portion of its arms, at whatever point they may have been severed; of the numerous specimens which have been scientifically examined, many showed that one, two, or more arms have been either repaired or reproduced; and some of the female specimens have shown a loss of the whole eight arms, but all more or less restored.

Another kind of exuviæ observed with the octopods is the outer skin of their long limbs, which they not infrequently shed. These cast-off skins float upon the water, and are one of the indications which lead to the discovery of their retreats. When the outer skin becomes too tight for the growing animal, or is worn too smooth by frequent contact with the rocks, the creature may be seen rubbing its arms against each other as if they were undergoing a scrubbing or cleansing process, and soon these thin, filmy skins may be seen floating away on the surface of the water.

At certain periods there appears in the male octopus what is called the *hectocotylus* development in one of the arms. When this gentleman would a-wooing go, as Mr. Lee says in his valuable little book on this subject, and " he offers his hand in marriage to a lady octopus, she accepts it most literally, *keeps it, and walks away with it ;* for this singular outgrowth is detached from the arm of the suitor, and becomes a separate living creature," specimens of which have been preserved in the Museum of Natural History in Paris. This *hectocotylized* arm is afterward reproduced in the male.

It is surprising with what care the female watches over the development of the eggs. Having selected a snug retreat in the rocks, she will barricade it by dragging to the entrance other portions of rock, or perhaps a pile of oysters —anything out of which she can make a strong breastwork or line of defense ; and then she sits on guard ready to attack any intruder, even though it be her own mate. The eggs when first laid are about the size of grains of rice, and are arranged upon a stalk which is attached to the rock by a cement secreted by the parent, and to which each egg is separately attached, like a mass of bananas on its stalk, only much more closely packed, the number being immense ; an octopus will produce in one laying from forty to fifty thousand. Mr. Lee describes one that he had under observation in an aquarium, which he says " would pass one of her arms beneath the hanging bunches of her eggs, and, dilating the membrane on each side of it into a boat-shaped hollow, would gather and hold them in, as in a trough or cradle. Then she would caress and gently rub them, occasionally turning toward them the mouth of her flexible exhalent and locomotor tube, which resembles the nozzle of a hose-pipe, and direct upon them a jet of water." The object of the syringing process was probably to free the eggs from parasites, or to prevent the growth of confervæ upon them. At the end of five weeks some of the eggs were taken from the nest for observation under the microscope, which showed that the young octopods were already alive and freely swimming within the shell ; and most extraordinary was it to see that these immature creatures exhibited the characteristic changes of color at that early stage of development, flushing red apparently with anger when disturbed. The period of incubation is about fifty days, and during all that time the mother octopus brooded her eggs with the tenderest care ; so that the observer almost ceased to look upon her in the light of a " devil-fish," and recognized that at least the maternal instinct was not dependent for its development upon external beauty.

When the young octopus emerges from the egg it is about the size of a large flea, but has none of the arms developed; these appear simply as "rudimentary conical excrescences, having points of hair-like fineness arranged in the form of an eight-rayed coronet upon the head." The amiable disposition of all female devil-fish is not perhaps equal to that of the one described above; but it is not an unusual

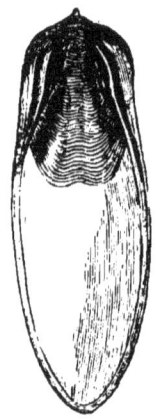

SEPIA OFFICINALIS AND SHELL.

event for them to die from the effects of exhaustion at the end of the long brooding period. This may perhaps partly result from insufficient nourishment, as they must evidently miss many chances of obtaining food, which others, unburdened with family cares, avail themselves of.

The nearest relations of the octopus are the cuttle-fish and squids. The former, *Sepia officinalis*, is best known as the animal which produces that fine black coloring fluid known as sepia-ink, and for its useful *sepiostaire* or internal shell, which is usually hung in the cages of canary-birds.

Though the cuttle-fish resembles in its general structure its relative the octopus, it varies in several particulars. Instead of eight arms, it has ten, eight short and two long. Some persons have pronounced them "beautiful"—in which opinion we cannot coincide; but their manners are decidedly more genial than those of the octopus. Instead of lurking in semi-concealed caves or behind rocks, and springing upon the unwary like a tiger from its jungle, the cuttle-fish comes out to the light and gives his intended victim a fair chance, having more the habits of a bird of prey than its congener the devil-fish. It is, however, very voracious, and fishermen have often cause to regret its proximity to the fishing-grounds, as it will attack fish while entangled in the nets and drag them out or bite and mutilate them. When attacked, its best defense is the sepia-bag, from which it ejects the black fluid, thus discoloring the water and escaping in the obscurity.

The eggs of the cuttle-fish are usually found attached to a branch of sea-weed and very ingeniously hung by a perfect loop, each one separately upon the twig, where together they somewhat resemble a bunch of grapes. As soon as the young are released, they seek the light and approach the surface of the water. The sepia is naturally very shy, and at the slightest alarm shoots forth with wonderful rapidity its foe-defying ink; but in captivity its fears may be overcome by kindness. It is not difficult to tame, and in time it appears to recognize and appreciate its protector, ceasing to discolor the water when sufficient familiarity has been established between them.

The eyes of the cuttle are so solid as to be almost calcareous, and are divided by a groove in the centre; these halves are nearly globose at their outer surfaces, and reflect light with a "beautiful nacreous opalescence and play of colors." In Italy they are made into beads for necklaces. The cuttle-bone when pounded is used as a polishing powder by jewelers, under the name of "pounce." It is also manufactured into a dentifrice, and sold under the name of

5

"white coral-powder." Artists still use the natural *sepia* to some extent.

The *common squid* (*Loligo vulgaris*) has the same num. ber of arms as the cuttle, but differs in form and some other particulars. The body of the cuttle is of a broad oval shape, with no perceptible neck; the squid is nearly triangular in shape, and has two plainly-defined necks, one much smaller inserted within the other and projecting beyond it. It has also very large eyes in proportion to its size. It is a free swimmer like the cuttle; its spawn is also left to float freely, but in a large circular mass, consisting of an immense num- ber of branches, all containing quantities of ova and united to a common centre. It has been estimated that these "mop- like" masses contain nearly forty thousand eggs. The squid is also privileged to carry an ink-bag, of which he makes very free use; and many fishermen attempting to catch them have experienced the fate of Tom Hood, of whom Mr. Lee tells the anecdote that, being unaware of this propensity of the cuttle-fish and squid, and having caught one of the former on his hook while angling in Love Harbor, he laid hold of it to unhook it, and received its full *jet d'eau* in the face. On being asked what he had on his line, he replied that he did not know exactly, but thought *he had caught a young garden engine!*

As these sorts of creatures are never eaten in this country, it may be news to some that they are very extensively used as food in many countries at the present time, and that the ancient as well as the modern Greeks considered them a deli- cacy when properly cooked. One cause of the favor in which they are held by the Orthodox Greek Catholics on the shores of the Ægean Sea is the substitute which they offer in place of meat and fish, both of which are forbidden during the long fasts of the Greek Church. A cuttle is practically de- clared not to be a fish, and certainly it is not meat; and so it finds its way into the pots and frying-pans even of the ecclesiastics during Lent and other fasts in great quantities. A common way of catching them in the Mediterranean is by planting traps of stone jars or earthenware tubes, into which they creep, and are thus drawn up and secured. Everywhere they are used for bait, and the Indians of Vancouver's Island and Alaska eat them with relish, as do the inhabitants of China and the western coast of South America. There is a good story told of a party of *savants* in England endeavoring to make a dish of one, at a special dinner given for the purpose; but the attempt was a complete failure—no one could swallow a morsel. The ancients described them under the name of *polypus*, and all classical scholars will recall the frequent references to these animals as articles of diet, especially by the comic poets.

The greatest enemies to the class of cephalopods are the porpoises, dolphins, and conger-eels. The last do not hesitate to attack even a devil-fish of considerable size, while the young are snapped up by a great variety of fishes. In fact, if the great mass of all the spawn produced by the denizens of the ocean were not devoured or otherwise destroyed, the watery world would long ago have become so over-populated as to be unnavigable, and its condition incompatible with the health of the human race.

CHAPTER VII.

MOLLUSKS.*

THE BORING PHOLAS, TEREDO, ETC.

THE immense variety of the mollusca forbids us to attempt the portrayal of more than two or three members of this division of marine animals. The name *mollusca* is from the Latin *mollis*, meaning soft, because all these creatures are soft-bodied. They are separated into two grand divisions, the shelled, called *conchifera*, and the naked or shell-less, *tunicata*, which conceal themselves in a sort of leathery tube. Next they are distinguished as *cephalous* or *acephalous*, that is, with visible heads, or without that apparently essential member; again as *cephalopoda*, those whose feet extend from their heads; *pteropoda*, those which have little wing-like expansions near the head; and *gasteropoda*, having a muscular foot extending from the under portion of the body. The shell-bearing mollusks are divided into *univalve*, *bivalve*, and *multivalve*, according as they have one, two, or more shell-plates; and, finally, into the *gregarious* and the *solitary*. These last are considered the veritable representatives of this immense class of marine animals, which includes almost every aquatic existence between a zoöphyte and a fish.

First, perhaps, it will be best to explain the process of shell-making; for, although the shell is apparently only the house in which the animal lives, it is equally true that if you turn him out-of-doors he will die. The connection between

---

¹ For illustration of the hermit-crab see page 14.

the inhabitant and its shell is a vital one. All these soft-bodied creatures are provided with a tough, leathery, tenacious sort of skin, the *corium*, which is commonly called its *mantle* from the looseness with which it covers the body; and it is from this mantle, which is a vital tissue, that is secreted the calcareous, earthy, horny, and sometimes glassy

OYSTERS, SHOWING DIFFERENT STAGES OF GROWTH.

matter which forms the shell. True, the embryo, while yet in the egg, shows a rudimentary shell, but this must necessarily be extended if it is to protect the growing and adult animal, which, unlike the crustacea is not allowed to cast off its old coat and procure a new one. This secretion from the mantle proceeds continually until the animal has attained its adult size, when it appears to be passive, unless called upon to repair injuries, when its activity is again apparent.

The curious varieties in shape which we see especially in the univalve shells, many of which are knobbed, ridged, or adorned with long spines, are all the result of the shape of the animal; every inequality upon the shell showing a ridge or protuberance upon the underlying skin; and the long spines equally indicate the projections which the animal has thrown out, like so many arms, within its mantle. In many shelled mollusks this mantle appears to have been interwoven

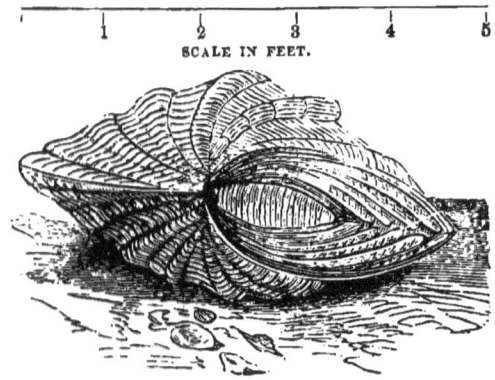

SCALE IN FEET.

GIANT CLAM (*Tridacna gigas*).

in a measure with the calcareous matter. This may be tested by submitting the shell to a strong heat, when the carbonate of lime will show white, and a fine black tracery will define the remains of the animal matter; or, test your shell by a weak solution of muriatic acid, when the earthy matter will be dissolved, while the organic, usually in flakes of albumen, will remain, a delicate framework of the original shape of the shell. The mollusk tribe vary in size from the microscopic to the gigantic. Some specimens of the *Tridacna gigas*, or giant clam, have been found nearly five feet across, and weighing over five hundred pounds!

Among the gregarious mollusca are the valuable edible bivalves—oysters, clams, mussels, and the pearl-bearing pinta-dines, and also the destructive barnacles. The *pholadidæ* and the *Teredo navalis* work singly, and it is to these *boring* mollusca that we shall give our attention in this chapter more particularly, partly for the reason that their habits are very interesting *per se*, and also because they have been the subject of much discussion among naturalists, and accurate knowledge in regard to them cannot fail to be serviceable to naval architects and all builders of docks, breakwaters, bridges, lighthouses, or any construction which may be exposed to their ravages.

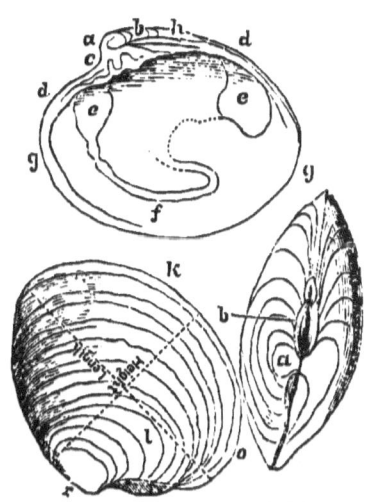

BIVALVE SHELL.—*a, a,* bosses; *b, b,* hinges; *c,* principal teeth; *d, d,* lateral teeth; *e, e,* muscular impressions; *f,* pallial impressions; *g, g,* sides of the shell; *h,* ligament; *k,* ventral edges; *o,* front edge; *r,* umbo.

The boring *Pholas* or date-shell is a bivalve of an oblong ovate form, which when closed is nearly cylindrical in shape, the upper or anterior part open; the shell is white and polished on the inside; the valves are not joined closely together, but are connected by a strong, tough membrane, to which is joined an extra plate taking the direction of the column of the main valves. It makes its home in either stone

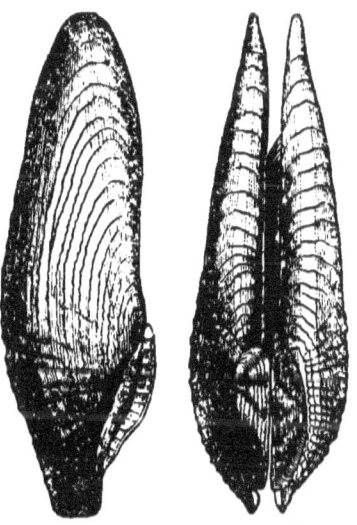

DATE-SHELL (*Pholas dactylus*).

or wood; in the latter it always bores across the grain. My specimens are snugly ensconced in a piece of rock, into which they are as nicely fitted as if the hole had been carefully prepared for them by some skilled mechanic *after taking their measure.* Rather better, indeed, for as the shell is larger just above the base than it is at the top, and the hole conforms to this shape, there is no artificial means by

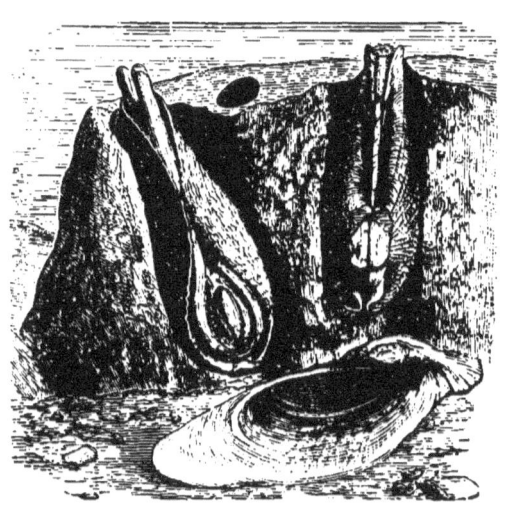

which the animal could be introduced to such a dwelling.

Many have been the theories, and long the discussions, as to how the animal drilled out this cavity. It was for long as much a mystery as the question was to King George, "How did the apple get inside of the dumpling?" But careful observation

STONE-BORERS (*Pholas dactylus*), which have hollowed out shelters in a block of gneiss.

of pholas in the aquarium has settled the fact that the animal begins its excavation when young. Selecting its position, it drills by regular movements with the corners of the hinges of its shell, being aided in keeping its position by the integument binding the valves together, and as it drills it ejects the *débris* through a siphon-like tube. During the early part of its life it is obliged to work at intervals to make room for its increasing size, its object being evidently to protect the whole of its shell. When its maturity is attained it can have no motive to continue to labor. In fact, it needs to be just below the surface of the rock or wood, and no more, for if it buried itself too deeply it would be unable to reach out for passing food. As it never leaves its position, it is entirely dependent upon what floats by it. When fully extended, the animal protrudes beyond the cavity a trifle farther than the length of its own shell. When it draws itself within, it is amply protected by its rocky fortress from most of its enemies; but starfish are exceedingly expert at extracting them from their holes, and in an aquarium must be closely watched or mischief may follow.

There is another little animal which is a stone-borer, and is usually found in caves and clefts of rocks subject to tidal changes; it is commonly known as "red-nose" (*Saxicava rugosa*). It prefers dark and secluded places, and is able to bore into the hardest limestone rocks. Its common name is derived from the circumstance that when extending itself from the orifice of its dwelling it presents a whitish, fleshy tentacle or tube-like proboscis, which is of a bright crimson at its extremity. This really consists of two parallel tubes, and out of it the animal squirts water if frightened or disturbed. Thus, if you attempt to pull one out of his domicile, and he perceives the intention in time, you will receive a few drops of water in your face, very well aimed. If you actually succeed in seizing upon his long, red nose, you are no nearer capturing him—*out you cannot drag him;* his proboscis slips through your fingers, in he pops, and there he sits, laughing you to scorn. Your only chance of securing him is not to dislodge him, but to take house and all, by breaking off the portion of rock in which he is intrenched; and, when he is placed in the aquarium, hunger will eventually compel him to show his crimson standard once more. His shell (bivalve) is of a dirty-white color, thick in proportion to its length, and rough on the outer surface.

A greater enemy to the ship-owner and dike-builders than even the date-shells or the barnacles can be is the *Teredo navalis* or ship-worm. This long, worm-like mollusk is all the more dangerous that it works so rapidly within the timber, while unseen from the outside ; at least, without a close and careful examination, its presence may remain unsuspected. Thus not only ships, but piers and dikes, are particularly endangered. The wood becomes perforated by millions of these unseen borers, who always work under

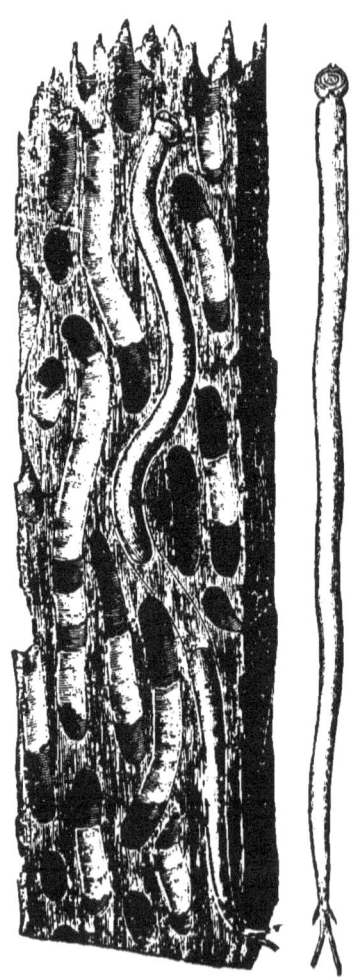

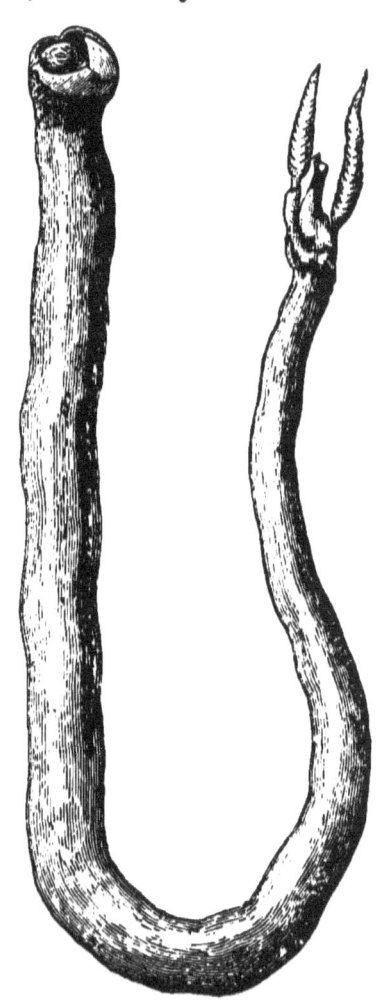

Wood exposed from November, 1874, to September, 1876, in crib at Pier No. 1, New York, North River, twenty-five feet below mean low tide.

This cut was made from a *Teredo navalis,* taken from a pile exposed two seasons (1876 and 1877) at Horn Island, Gulf of Mexico. When first taken from the wood it was eighteen inches long.

water; and, until we learned to copper-sheath our ships, doubtless many lives were sacrificed through the mischievous activity of these vegetarian mollusks.

The teredos occupy a middle position between the naked mollusks and the bivalves. The globose portion of the animal is inclosed in two small, thin, delicate valves of a shelly substance. Its mantle forms a fleshy sheath to its long, worm-like body, which is compounded of a double tube, joined for three-fourths of its length, but toward the extremity dividing into two separate tubes; or, we may say, it possesses a bifurcated tail. The use of the two tubes is to introduce through the one the organic particles which constitute its food and aërated water, while through the other are ejected the vitiated air and other exuviæ of the system. The exceeding narrowness of the body requires that all the internal organs should be placed as it were in procession— one following the other in a line.

The young larva, as soon as it leaves the egg, is able to swim about by the aid of vibratory *cilia*. Its immediate object in life is to find a piece of wood which it may begin to bore; having found this essential to its existence, it may be seen *prospecting for a bore*, traversing the timber like an intelligent miner to decide upon an opening. Its incision, when the spot is finally decided upon, is in the direction of the grain of the wood (the opposite of the pholas). When it has effected an entrance sufficiently large to contain half of its body, it commences the secretion of its shell, first exuding a mucous fluid which soon hardens to the necessary consistency. In about three days the process is completed.

In this shelly covering holes are left through which the tubes or siphons may be projected, and which are lengthened or contracted at will. The *boring apparatus*, in the case of the teredo, is believed to be the stout cutaneous folds which envelop the anterior portion of the animal, which are covered with a thick, leathery epidermis, and moved by four strong muscles. It cannot be the shell, for that is not formed till the work is nearly done. Some naturalists assert that the young teredo *feeds upon its own chips*, that is, upon the raspings of the wood while it is making the perforation. This, however, needs confirmation. Dr. E. H. von Baumhauer says:

"The teredo does not always remain in peaceable enjoyment of the home he has constructed, and the nourishment the water

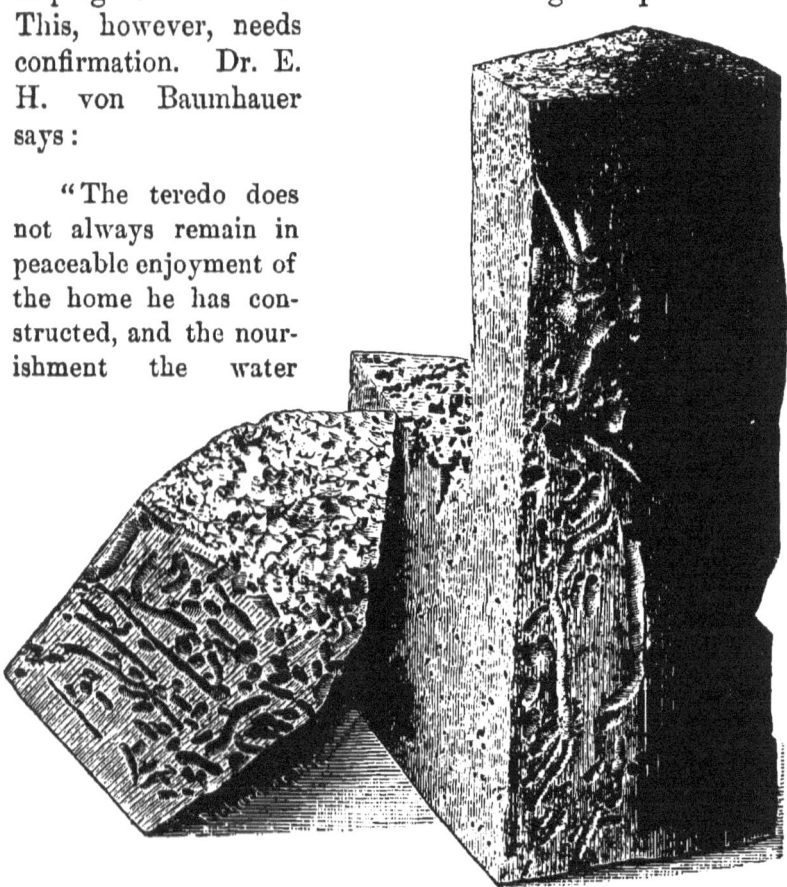

This cut was made from a piece of pine-slab, partially creosoted and exposed one season (1877) in the Gulf of Mexico. Only the dark-colored portion on the right side of the block was creosoted for this experiment. The rest of the block, *untreated*, was entirely riddled by the *Teredo navalis*, up to the edge of the creosoted portion, but that the destructive marine worm carefully avoided.

brings to him. He finds himself exposed to the attacks of an enemy, of an annelid to which the late M. W. de Haan has given the name of *Lycoris fucata*. In our day, as well as at former epochs, this annelid is constantly found wherever the teredo exists. His eggs and embryos are met with in the midst of those of that mollusk.

"Kater has remarked that the adult annelid, leaving the muddy bottom, where he has hibernated, and in which the piles are driven, climbs along the surface of the wood toward the opening made by the teredo ; there he sucks away the life and substance of his victim ; then, slightly enlarging the aperture, he penetrates and lodges in place of the teredo. Later the annelid reappears and seeks for new prey. All the early writers on this subject state that they have found this annelid in wood at the same time with the teredo. It is remarkable that a similar annelid, and perhaps the same, has been found in the cavities hollowed out in stone by the *pholades*.

"It is important that it should be generally understood that this annelid is not only harmless, but renders the greatest service in devouring the wood-destroyer. It is a narrow annelid, ten to fifteen centimetres long, provided on his sides with a great number of small feet terminated with a point and garnished with hairs, and showing in front a pair of strong upper jaws, horny and sharp, and lower jaws bent backward in form of hooks and carried outside by the aid of the lower lip, which is developed somewhat like the finger of a glove turned backward. Behind the head are four pairs of tubular-formed gills. With these weapons the annelid pursues and devours the teredo. The observations of Kater teach us that he is generally found in the empty galleries with the remains of the teredo ; sometimes even he is seen as if clothed with the integuments of the teredo, while he is occupied in ransacking his intestines."

LYCORIS FUCATA.

Some of the commonest bivalves found upon our shores are as interesting as the more scarce varieties, if one will bestow sufficient attention upon them. There is one with a somewhat sanguinary name, the "bloody-clam," or arca

(*Argina pexata*), which deserves a brief notice. Its name of *arca* is derived from its supposed resemblance to a chest or box; but to our eyes it must be a chest with the corners knocked off, for it has no acute angles. The outer surface is roughly corrugated, and covered with hair-like spines. But its principal attraction is its bright, orange-colored mantle, which ornaments the edge of the shell in ample folds like a gorgeous Elizabethan ruff. Its common name the fishermen have applied to it from the fact that when they open it a red, sanguinous-colored fluid is discharged. This animal has the habit of tying itself down with a byssus, as is the case with the mussels and many other mollusks. I have kept this species for a long time in my aquarium. They are peaceable, and have no offensive habits. Indeed, this is true of all the mollusks which anchor themselves in the manner described; they accept what prey Providence sends them, but of course they pursue none.

Mytilus edulis, or edible mussel, is the small, black-shelled mussel commonly sold in the markets, and sometimes preserved by pickling. In France and some other parts of Europe these mussels are cultivated very extensively, and furnish food to the inhabitants in even greater quantities than does the oyster. They are thickly strewed on our shore in many localities. At Fort Hamilton scarcely a stone can be found at low tide but is covered with them; and even old wooden posts standing in the water a little distance from the shore are loaded with them. These belong most emphatically to the class of gregarious mollusks, occurring often in heaps, so that they have been used in some instances as natural buttresses to support bridges or breakwaters near the seashore. The combined strength of bushels of mussels, all fastened together with their silken byssus-threads, which are very strong, makes a more resisting wall than can be built by the human mason.

On our own coasts they live and flourish, simply opening

their valves when hungry to receive whatever the ocean-waves may bring them. Millions on millions are, however, annually destroyed by the blackfish (*Tautog*), which is very fond of nipping them off, especially when they are young and their shells are comparatively soft and tender. Some of these shells are very prettily marked with light, striated bands; the lines form a sweep something like the tail of a comet, taking the narrow, hinged portion of the shell for the head or nucleus of the stream-ers. Occasionally we observe among the mass of jetty-black shells one or more of a deep orange - color. In the tank their most attractive feature is the perfect byssus which they spin from an internal organ, which only the fisherman sees in its natural state, and which is usually a great novelty to the inland amateur.

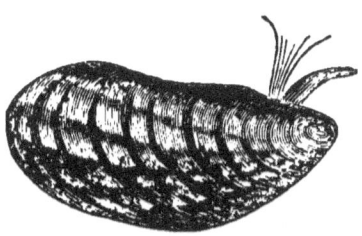

COMMON SALT-WATER MUSSEL (*Mytilus edulis*).

There is another variety of the mussel, not edible, called the *Modiola plicatula*, and which is equally plentiful as the last on our shores. It has a very handsome shell, corrugated and marked with mathematical perfection over its entire sur-face. It is used extensively for ornamental purposes, the rough outer corrugations being ground down to a perfectly smooth surface, and then polished with rotten-stone and rouge. All the beauty of the corrugated markings remains visible through the substance of the mother-of-pearl which is revealed in the process; or, to speak more accurately, the nice shading of the original undulations preserves the *ap-pearance* of these markings, while the surface is really as smooth as glass, and shows beautiful iridescent colors. Some of the flats on and near Coney Island are literally paved with these shells so thickly that one can walk over them for miles;

and little do many people think, as they crush them, that in so doing they are trampling under foot the rudimentary jewelry they may some day pay a high price for the privilege of wearing.

Modiola modiolus is a species which I have never found nearer to New York than the Massachusetts coast, though I believe it exists here. It is rather an interesting animal for the aquarium. It is very hardy and gives no trouble. Its shell is covered with soft spines or hair-like appendages, for which I have discovered no use, unless when in its natural state they serve to conceal the animal from some of the voracious fishes which are constantly preying upon the mussel tribe. They display no sociability in the aquarium, neither troubling their neighbors nor me; in fact, though I rather like the little creature, which I have had for more than a year, I have never yet been able to elicit from it a glance of affection or even recognition !

Mya arenaria.—Did anybody ever think of the prosaic clam as an object of scientific interest—the veritable " soft clam," which is a staple article in our markets ? Despise it

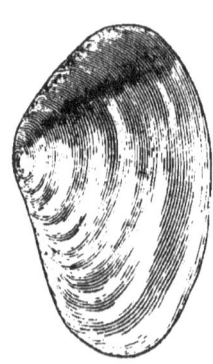

not; it has its virtues, besides those succulent ones recognized by the connoisseurs of the table. Writers on the aquarium usually pass this animal by, or content themselves with remarking that it will not thrive in the aquarium. My experience does not coincide with this. I have had one in a healthy, lively condition for a long time, and have found it of quite sufficient interest to pay for the room it occupies. Its manner of sustenance is provided for by the simple arrangement of a

SOFT CLAM (*Mya arenaria*).

double tube through which the food-bearing current enters. These tubes are furnished with a delicate, pretty, white fringe,

which acts as a kind of filter, preventing objects too large
to be received into the stomach from entering. Under a lens
this gentle current is easily detected by the flow of particles
in the water, which may be seen entering the tubes.

The *Venus mercenaria*, or common hard-shell clam, is
too well known to need description here. Its shell is large,
round, and of stony hardness. It will
thrive in the aquarium, and is not so
destitute of good looks as the unobserv-
ant might imagine. When open it
shows a very handsome double frilling
of pure white. It is called by the Long
Island fishermen the *quahog* clam.

HARD-SHELL CLAM (*Venus mercenaria*).

Anomia glabra, or the jingle-shell,
is one of the most numerous and widely distributed mollusks, and I venture to say that this pretty little shell has
been picked up by almost every person who has ever visited
the sea-shore. They are sometimes called "silver-shells" or
"golden-shells," because some of them have a silvery appearance, while others have a yellowish tint; they are very
thin, even fragile in texture, and not usually larger than a
nickel cent. When a number of them are assembled together, there may often be heard a sort of jingling sound.
But the prettiest thing about this shell is to possess it when
alive, and observe the proceedings of its owner. I always
have more or less of them in my vases, for they are as harmless and modest as they are attractive in appearance; and I
should not consider my collection as in any way complete
without them.

They are eminently gregarious in their habits, and have
a most singular way of arranging themselves in layers and
bunches. A favorite mode with them is to select an old
scallop-shell (*pecten*), and arrange themselves around it as
far as the space permits, on the concave side; and, *if there*

are more of them than will go round, they pile up layer upon layer, one on the top of the other. But this is not the strangest part of their proceedings; the way in which they fasten themselves to the old shell, or to each other, is a novelty even among the many wonders of the lower orders of creation. Their mode is to *rivet themselves* to whatever object they desire to attach themselves to. If you find a clump of these shells and attempt to separate them one from the other, and from the base on which the group is fixed, you will find that each shell on its under part is perforated, and into this perforation a little excrescence is inserted and nicely fitted, just as a rivet would fasten one piece of iron to another. This is the mystery of the holes which are so often observed in these little silver and golden shells.

It is curious to see, if one gives the group a slight tap, or if they are in any way disturbed, how each one closes its doors in regular succession, as if an inaudible order had been given, "Close shells! *one—two—three!*" and so on, till the whole mass is as thoroughly protected from intrusion as their delicate coverings will permit.

The little boat-shell (*Crepidula fornicata*), when emptied of its inhabitant, bears a close resemblance to a row-boat, with a very serviceable seat in the stern, extending quite across, with a hollow space beneath it. These mollusks are mostly found on oyster, clam, or other shells; sometimes on stone or wood, or on the coarser kinds of algæ. Their general appearance in this position is like a quantity of oval-shaped, shallow cups piled together in a reversed position, for they pile themselves up in tiers five or six deep. To see the animal which inhabits them, we must pull the shells apart, and look on the under side.

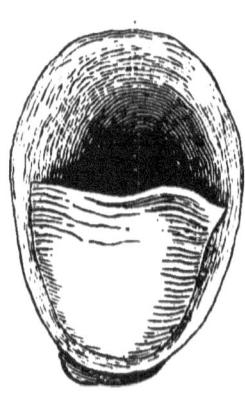

BOAT-SHELL (*Crepidula*).

But we might go on almost endlessly without being able to give, after all, more than a fraction of this numerous—we might say inexhaustible—family of mollusks. Fearing to weary the reader, we shall add but one more to our list.

The *Chiton* (called also *Chætopleura*) is a gasteropod, attractive from its handsome and serviceable shell. The name

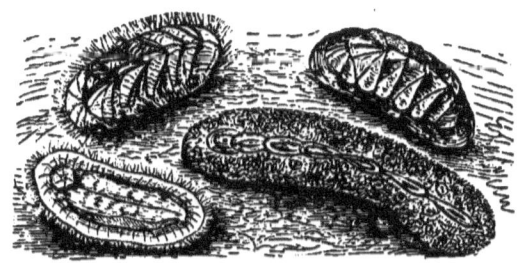

CHITON SQUAMOSUS AND C. SPINOSUS.

chiton, that of the ancient Greek cuirass or coat of mail, was given it from a fancied resemblance in the folds or joints of its carapace to that article. To our eyes it is strongly suggestive of the "painted water-turtle" (*Chrysemys picta*), for there is a striking analogy in the markings on the shell of this reptile to the imbricated carapace of the chiton. But its structure is entirely different. Instead of consisting of one solid piece, our little mollusk is arrayed in chain-armor, the eight bands of plates overlapping each other, and united by an integumentary substance which enables the animal to roll itself together like a wood-louse. The chiton grows to a length of two and a half to three inches, and is of a greenish-gray color, the divisions of the shell transversely being marked by darker lines, something like shagreen skin. It has a suboval foot, and can if it chooses cement itself firmly to a shell, or any hard substance, by a gelatinous fluid which it secretes. Its head consists of a puckered frill of a membranous consistency. It is furnished with certain small or-

gans resembling the spiricula of the annulose tribe—two
little wheel-like objects which, working in reverse directions,
act like the grinders of vertebrate animals in preparing food
for its stomach.

The chitons are very widely dispersed, being found in
almost every part of the world, some varieties remaining always in deep water, while others near the shore have the
habit of sunning themselves on the rocks.

Botryllus violaceous.—The *Botryllidia,* sometimes called
" little bottles " from their shape, are a curious combination
of partly gelatinous and cartilaginous objects, whose exact
status it is difficult to fix ; for some of them have also tiny
shells of a very delicate nature, while their general appearance to the naked eye is more like patches of bright-colored
jelly than anything else. Sometimes they are arranged in
the form of a bunch of grapes, and their color is of a greenish gray with dark-blue markings ; again they may be found
of a white color with black star-like centres, or the reverse ;
the colors in other groups will vary from a light yellow to a
deep wine color. The shelly substance is of the nature of
spiculæ. They are mostly attached to rock or stones subject
to tidal changes. Each of the little animals maintains a
nearly vertical position, and the more closely they are inspected the better do they repay investigation. They seldom
exceed over a fourth of an inch in length, and are usually
aggregated in patches of five or six inches in diameter.
They belong to the tribe of ascidians. Gosse places them
among the mollusks. Quantities of the *Botryllidia* may be
found among the rocks at the northern end of Manhattan
Island, along the shores of the East River and Long Island
Sound. The searcher for these interesting groups will
usually find them arranged in a nearly circular form, or
rayed like the spokes of a wheel ; and these show a sensitive-
ness of the whole community if the centre, or what we might

call the " hub," is touched, while if the extremity of one of the rays is disturbed, only the individuals annoyed will evince sensation.   They are a kind of link between animals possessing a purely individual life and that of the hydroids, which is essentially compound.

CHAPTER VIII.

THE HERMIT-CRAB.

" We shall be well off here," said I to Conseil. "As well, by your honor's leave, as a hermit-crab in the shell of a whelk," said Conseil.— *Jules Verne.*

THE hermit-crab, also called the soldier-crab (represented, with a sea-anemone on its back, on page 14), is truly the step-child of Dame Nature, the homeless *gamin* of the shore ; "being sent into this breathing world," as Richard of Gloucester says, "scarce half made up." In other words, while its anterior portion is provided with the shelly vesture of other crustaceans, its rear portion is left without any protection. Hence may have been developed those pugnacious qualities which have added to its title the warlike appellation of "soldier." For necessarily the animal becomes *a soldier of fortune*, the first business of its life being to find, seize, hold, and defend some shell into which it may thrust the soft part of its body, and thus save it from the attacks of enemies and the voracious bites of hungry fish.

The American hermit-crab, the *Pagurus longicarpus*, is distinguished, as its name imports, by the length of its (right) hand or claw. The English variety, the *P. Bernhardus*, is the species described in most of the popular "sea-shore books." But in the State of New York, the Carolinas, and Georgia, including the sea-coast and fluviatile species, many varieties have been recognized. They are usually supplied with shorter antennæ than other crabs, and the soft end of

their unprotected extremity is furnished with a pair of un-
equal appendices, the use of which we shall describe as we
proceed. It is not easy to find perfect adult specimens of
the hermit-crab, many of them having lost their antennæ and
one or both eyes. The fishermen attribute this to the attacks of the blackfish, which are believed to bite them off;
but the fact may equally well be accounted for by the pugnacious habits of the creatures, who frequently indulge in
sparring-matches among themselves, especially when a new
residence is an object of desire to more than one. Some
form of the whelk-shell is usually preferred by the hermit.
The young use a shell suitable to their size until it becomes
too small for them, when they set out in quest of something
better adapted to their growth. If empty shells are plenty
along the shore, the work of selection goes on carefully, even
warily, but without difficulty. The hermit invariably puts
his long claw into the supposed empty shell first, to see that
there is no concealed, dormant inhabitant within; having
made sure of this, he whisks his hinder portions into it with
marvelous rapidity, keeping hold of his old shell in case he
should not like the new. Sometimes he will try several
shells in succession, examining and testing them as carefully
as a dandy will try on numerous coats before he gets one
just to his mind. Having finally selected his shell, he remains in happy possession, unless he should unfortunately
be attacked by a brother hermit who thinks he would prefer
that particular dwelling to his own. If of equal size, the
hermits are cautious about coming to blows, but will stand
in something like the attitude in which Russia and England
stand to-day (April, 1878), glaring at each other without
either daring to take the initiative; but if the aggressor is
sure that he is the stronger of the two, the other may find
himself pulled out of his shell, the victor taking immediate
possession, and usually making a fine repast upon the van-
quished. If in an aquarium two shell-less hermits are put

into a tank with only one shell which can be occupied, then comes the tug of war; each will endeavor to seize it, and the " best man wins." Among no class of animals can the doctrine of the " survival of the fittest " be better exemplified ; with scarcely any do the absolute qualities of strength and courage find such constant occasion for use. Two tigers may fight, and the beaten may retire to his own jungle and nurse his wounds; two hawks may make the feathers fly in a prolonged contest, but the world is wide, and the worsted party knows where to seek a retreat ; two common crabs may come into collision and despoil each other of a claw or so, but their claws will grow again ; but our poor hermit, if dragged out of his shell, is not only houseless and homeless, but is, like Cardinal Wolsey, " left naked to his enemies." No one knows better than the animal himself the risks of this condition, and until he can refit himself with a dwelling he goes about as disconsolate as a New-Yorker might do, who had been turned out of his house, and found himself approaching the first hours of May-day with no new residence secured.

It is not always necessity, however, which forces a hermit to change his residence. He may have a sufficiently commodious and comfortable house, but, if he meets with one which strikes his fancy better, he has no sentimental regrets or reminiscences over the old homestead ; but out he goes from the old and into the new with the facility of a Western pioneer when " prospecting " for a location. Probably the choice is really determined by the capacity of concealment and consequent protection which a given shell offers ; for the crab, in fitting himself into his home, dexterously clasps the most deeply-recessed whorl of the shell by the two hooked appendices of his caudal extremity, and admirably adapts the soft part of his body to the interior ; and if the shell will allow him to draw a portion of his anterior parts under shelter, so much the better. It is astonishing what force he exerts to retain his protecting shell. It is no easy job to get him

out either by force or stratagem ; and, unless when desirous
of making a change himself, he never leaves his shell except
when sick, or suffering from impure water in a tank, or approaching death, when he always vacates his borrowed house,
as if he did not wish to be found *in extremis* with stolen
goods upon his person. In the aquarium his motions in this
respect will plainly indicate if there is any fault in the water ;
if it has become impure from any decayed substance, or
supercharged with carbonic acid, our friend will quit the
shell he may have been at much trouble to obtain, and climb
up toward the surface, on the rock elevation or a growth of
algæ.

In life the hermit-crab assumes many characters. In his
assorting and choice of old deserted shells, he may be considered as a *dealer in old clothes*—a unique profession in
animal life ; for though some birds appropriate the nests of
others, and bees will occupy artificial homes, yet for a personal vesture the hermit-crab alone is dependent upon other
animals. Nor is he unsuspected of "flat burglary as ever
was committed," and sometimes murder withal ; for he has
been known to assault a disabled whelk that was unable to
defend itself, put in his long claw, draw out the poor animal,
and slip himself into the yet warm dwelling, and march
about with his ill-gotten prize, as proud as the most noble
warrior-robber in the world. Yet for all his sins is not Nature accountable, in having thrust him into the midst of
enemies so indifferently accoutred ? But must we not add
hypocrisy to the list of his vices, in that he always appropri-
ates the shell of an animal which is meekness itself, and thus
deludes those that would prey upon him into the belief that
they would meet with little resistance ? Besides the whelk-
shells, the *Fulgur carica* is his choice.

But with this long and serious presentment against him
there is much to be said in his favor. Give him a respect-
able shell, and enough to eat, either mollusks or meat, and

6

he will usually be harmless and peaceable enough. He is also very active, and causes much amusement by his constant efforts to scale every piece of rock or other elevation, such as the stronger vegetable growths which can bear his weight. When he happens to fancy a slippery frond, it is curious to see how he will persevere, " hauling, slipping, and tugging," to get to the top; and if after immense toil he reaches the topmost pinnacle of a piece of rock-work, where he is very apt to lose his balance and fall plump down again, he takes it all as part of the play, and repeats the experiment, or seeks some new scene for his exploits.

If the hermit is obliged to take up with a damaged shell, in which he is not securely lodged, it is quite pitiful to see the efforts which he makes to retain his hold ; for if he has not a firm grip upon the interior whorl of the shell he is inhabiting, he is not only exposed to the attacks of his natural enemies, but is unable to resist the action of the waves, and finds himself tumbled over and over in spite of his desperate efforts to grasp at stones or anything within reach by which he may obtain a purchase to prevent being swept away.

The peculiar formation of the hermit-crab is well adapted to the habits of the animal. The claws in particular are so modified that the longer one serves for a sort of barricade at the shell's mouth, while the smaller one is easily tucked away within, taking up but little room. The curious kinds of companionship which sometimes exist between very different animals—such as the pilot-fish with the shark, for instance—are exemplified in the frequent presence of a small worm, known as the *Nereis,* in the shell occupied by a hermit-crab ; and it is highly amusing to see this secondary intruder suddenly slip out and snatch away some choice morsel which Mr. Crab was about to place in his own maw, and the different degrees of patience or vexation which the hermit will exhibit at his loss, sometimes allowing the theft to go unnoticed, and then again trying to recover it by thrusting one of his claws into the interior of the shell. Besides the occasional loss of a tidbit in this manner, the nereis is often an indirect cause of calamity to the crab; for this little worm is much valued for bait, and fishermen often turn the crab out of his shell simply to get at the worm.

These hermit-crabs may often be found in the tide-pools left upon the shore anywhere in the vicinity of New York, on the islands in the bay, or on the shores of Long Island; and small ones may sometimes be obtained at the fish-markets, among the silver shrimp snugly ensconced in whelk-shells.

Besides the intrusive nereis, the hermit is often attended by a parasite actinia, which, however, is satisfied with a seat upon the roof of this ambulatory house. Mr. Gosse, the famous English naturalist, believes that there is sometimes a conscious companionship between the actinian *Adamsia palliata* and the hermit; and he narrates an incident to prove it of which he was an eye-witness. He says that the crab, having chosen a new home, returned to the old one, and with his claws carefully detached the actinia from the abandoned shell, and placed it upon the top of the new—actually giving the zoöphyte several taps with his great claw, as though he would say, "There, my friend, be quick now and attach yourself!"

Another species of actinia is too loving to be wholesome. This is the "mantle anemone," which entirely clasps and eventually grows over the hermit, destroying its life; and reciprocally the anemone appears to die of grief, for it does not long survive.

If a hermit-crab cannot find a shell of any description, he will temporarily seek shelter for his exposed body in a hole in a rock, or even in a sponge; and one in captivity, having no other resource, took up his residence in a living actinia, dragging the poor thing about with him wherever he went.

Strange to say, this companionship endured for a consider-able time without fatal injury to either party.

The mode of progression with the hermit-crab often looks clumsy and even painful, if he happens to occupy a shell too heavy for his size; but it will be remembered that he has a good grip upon the inside, and that his arms or claws are very strong. With these he grasps some inequality in the rock, or digs them into the sand, and then drags the shell after him; and it is surprising how quickly he gets over the ground with his voluntary burden.

How early in life the young crab commences "house-hunting" appears to be undetermined; but it is probably very soon after hatching, since there would never be a mo-ment's security for him while unprotected.

CHAPTER IX.

SOME CURIOUS DENIZENS OF THE SEA.

THE SEA-HORSE.—The name of this unique production of the marine world would scarcely suggest to the imagina-

THE SEA-HORSE (*Hippocampus Hudsonius*), feeding on Serpula.

tion anything like the reality. We should naturally suppose it was a large animal, instead of which it rarely attains a length of six inches; but though so different from the quadruped after which it is named, both in size and structure, there is something about the head and neck, particularly in the pose of the latter, which invariably suggests an equine resemblance.

Some of the ancient philosophers who appreciated the productiveness of the sea, regarding it, according to the theogony of Greece, as the mother of all living beings (hence Venus is always represented as having originally risen from the sea), knew of one variety at least of this little fish, for they gave it the name of *Hippocampus*, from two Greek words meaning curved horse. They believed it to be the embryo of Neptune's fiery steeds. American naturalists have given the name *H. Hudsonius* to a species found in the tide-water of the Hudson River. The genus is very widely distributed, being found in all seas.

Viewed superficially, the undoubted likeness to the horse is but a reminder to the thoughtful observer that a resemblance may be very striking, and yet have no basis in structural similarity; that a likeness, however marked, may be, and often is, only imaginary or analogical. In this mistaking of semblance for reality, and reasoning from mere form to function, lies the foundation of many popular errors; though there is no danger of any modern mistaking the sea-horse for the progenitor of a quadruped. Still, as great errors as this would be have actually misled many observers in the field of natural science, particularly in the matter of mistaking the young of certain marine animals for distinct species, and classing them according to some fancied resemblance with other animals with which they had no true affinity.

The parts of the sea-horse which most resemble the real horse are totally different in internal structure and function from the supposed prototype. Following the outline of the little fish, we observe that the part of the head which looks like the jaw of a horse is in reality the respiratory apparatus or gills of the fish. The nostrils, instead of being at the end of the snout-like projection of the head, are really close below the eyes, and the mouth has a valvular form, and, though in the same situation, opens upward or vertically, instead of longitudinally. On the upper part of the head, on either side, are two little organs bearing a resemblance to ears; and along the back is a similar structure which might be likened to a mane, but which really consists of fins, the only organs of progression which enable the animal to swim about, since it has no caudal fin; all which similarity of appearance, with dissimilarity of function, illustrates the absurdity of arguing from analogy to identity. The office of true science, with its keen eye, is to look through form to substance, through appearances to reality; and in doing this multitudes of old errors are constantly being discovered and swept away.

If we are asked what the sea-horse really is, we answer, *a fish, in a measure*—not wholly; for while he is in many points fish-like, he is yet totally unlike any other fish in an important particular. The grand division of fishes is into two classes. In the first of these the backbone is prolonged into the tail, and one lobe of the caudal fin is much larger than the other; hence this class is called *heterocercal* or unequal-lobed. These are now few in number, but abounded in the early ages of the globe, when fish-life was at its maximum. Most of the present representatives belong to the *homocercal* class, those having tails with equal lobes. There is a great diversity in the shape of fishes' tails, which may be considered their rudder or steering apparatus. But our poor little sea-horse cannot be admitted into either of these two grand divisions; in fact, he has *no proper tail for a fish!*

In all other fishes it is the caudal appendage which determines the onward movement, and is capable of lateral motion only, striking right and left, as the water is turned aside by the starboard and larboard turn of a ship's rudder, or like an oar in sculling, which equally propels the boat onward.

But while the sea-horse has the same general osseous and muscular structure as ordinary fishes, its tail, instead of ending
in a fin, is prolonged far backward, and is capable of free
vertical movement, and is prehensile as a ring-tailed monkey's
caudal extension. It is not in any sense an organ of progression, and is usually carried in a curve or coil directed
downward, and (quite contrary to the rule in ordinary fishes)
has very feeble muscles for lateral motion.

Not all the fins of fishes are employed in aid of locomotion. Some are for the purpose of accurately balancing the
fish; and if one of these is cut off or injured, the fish there-
after swims unevenly—lop-sided. Should both of the pair
near the gills be removed the poor fish would have to swim
with its head pointed downward; the body could no longer
retain its horizontal position, and the *fish would be in danger
of drowning* from its inability to aërate its gills. There is,
however, one fish, of the genus *Monoptera*, which has but
one fin, and that the caudal—a complete contrast to the sea-
horse.

The absence of posterior limbs in an animal is indicative
of a low position in the scale of Nature's productions; for
instance, some lizards are thus "degraded," as the scientists
term the lower forms of a class. The sea-horse, however interesting he may be to us, must be placed in the same cate-
gory; for, alas! he has only *one pair of fins*, and these are
on the sides of the head, and he has but a single fin upon the
back. Nevertheless, all unconscious of his scientific "degra-
dation," he manages to enjoy his life and to move about quite
rapidly by the aid of these deficient propellers, with an even,
gliding sort of motion, the fins maintaining the while a rapid
vibratory action. The principal one, which is the dorsal,
has exactly the screw-like motion of a steamer's propeller.
Owing to the limited number and peculiar position of its
fins, it swims with its body in a perpendicular position. The
motion of the fins is in its character analogous to that of the

cilia of polyps and others of the lowest orders of animals, as the rotifera.

The prehensile tail is used to clasp around objects like a finger; with it, it will seize hold of almost anything and support itself thus attached to sea-weeds or even floating substances. I have had some in my aquarium which became so thoroughly domesticated, that they would answer my signal of tapping on the glass, come to the surface, and on my presenting a finger would coil the tail around it, looking inquiringly into my face with their bright, movable eyes. And here again it differs from ordinary fishes: it has the power of moving its head and eyes independently of its body, a power which is possessed by but one other fish of the present day, the gar-pike of our Western rivers. This variation from ordinary types suggests how we sometimes find the application of different organs to similar purposes : the sea-horse uses his tail for the same purpose as the elephant uses his nose (for the proboscis is but a prolonged nose), that is, to lay hold of objects, just as the wings of a butterfly do the service for it which fins perform for the fish.

The young of the sea-horse, like those of most marine animals, are produced from eggs ; but its mode of protecting them bears more analogy to that of the marsupial quadrupeds than of its neighbors the fishes. When newly hatched they are carried in a double fold of skin upon the abdomen, where they remain, like the young of the kangaroo in the maternal pouch, until they reach a sufficient degree of maturity to be trusted to go alone. The most singular part of this arrangement is, that it is not the female but the male sea-horse which thus protects and cares for the young ; being in this a counterpart to the little fresh-water stickleback, the male of which assumes nearly the whole care of guarding his precious progeny. But while the stickleback only watches the nest, the male sea-horse may be seen during any breeding-season with this external pouch distended with young. Unlike the young

of the kangaroo, the youthful sea-horse when once emanci-
pated does not return to the parental pouch, but maintains a
free and untrammeled existence.

A *family of young sea-horses*, swimming in a glass tank
of pure sea-water, is one of those rare sights which one may
be happy if he sees once in a lifetime, and which he will not
be likely to forget while life lasts. I once had a tank full of
them. The whole number I will not attempt to estimate;
but the water was clouded with their minute black forms,
and I suppose that every square inch of water contained at
least *one hundred living sea-horses*, each little head and neck
imitating the free movements of a horse almost perfectly. I
could scarcely reconcile myself to the loss of these unique
pets, which was caused apparently by their inability to find
their natural and proper food in the tank. Sea-horses, though
such interesting occupants of the aquarium, are not good
subjects for captivity. They rarely live long, and I believe
there is no recorded example of their successful breeding
when removed from their natural habitat.

The nearest approach to resemblance to the sea-horse is
found in the *Pegasus volans*, or flying-horse, a species of
flying-fish, the head of which bears some likeness to the sea-
horse. Its tail is also considerably prolonged, but it has a
small caudal fin, and is otherwise more closely allied to ordinary fishes.

THE TOAD-FISH (*Batrachus tau*).—If seen for the first
time on the shore, it might puzzle the observer to guess
whether this odd-looking creature was really a toad—"with
variations"—or a fish. It is usually of an olive-color,
mottled with green, and rather unsightly, as our illustration shows. It is nearly all head, and its countenance
is far from prepossessing; its flesh, however, is said to be
of a fine, delicate flavor. It is found in all our waters of
the Atlantic coast from Maine to the Gulf of Mexico. The
female of this fish is very careful of its eggs, depositing them

upon stones and sticks indeed, but cementing them securely
one by one in regular order, so that they may not be washed
away; they are very large, as compared with the eggs of
most fish, being about half the size of a pea; they are semi-

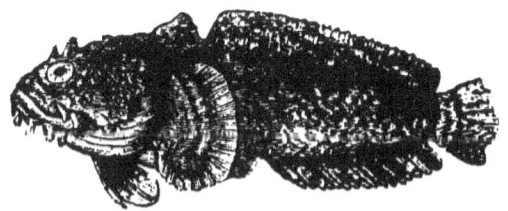

TOAD-FISH (*Batrachus tau*).

transparent, and are watched over by the parent until hatched.
The curious name of *tau* (which is the Greek *T*) was given
to it by Linnæus on account of a figure resembling that letter being formed on the head of a dried specimen by two
elevated lines.

The SEA-RAVEN (*Hemitripterus Americanus*) is another
large-headed fish, but much more ornamental than the toad-
fish or the angler. It has the appearance of having just
come out of a fight, its fins looking jagged and torn and its
face disfigured; but, to compensate for these detrimental appearances, its skin is of a rich, velvety texture, with blotches
of white and brown here and there, reminding one strongly
of some of those astonishing Japanese works of art in the
shape of dragons; and like these latter the sea-raven is *principally mouth!*—like the " end-man " in a negro minstrel
show. Though the written description of this fish would not
mark it as an attractive object, in reality there is much of interest about it, and it forms a very welcome addition to the
aquarium; especially as it is quite hardy, and has not that
" habit of dying " which is so annoying to the collector.

THE ANGLER (*Lophius piscatorius*).—This is a villainous-
looking fish, about four feet long, with a large head and
monstrous mouth. For some reason it has been blessed with

a greater number of names than usually falls to the lot of
fishes: in some localities it is called the "angler" or the "fishing frog," in others the "sea-devil," the "bellows," "goose,"
"monk-fish," etc., etc. Its habits are no better than its
sobriquets ; it usually establishes itself in an ambuscade of
mud or sand, and when the unwary little fish see the tempting bit of shining membrane which hangs over its mouth
from the tip of a curious long spine—like the "shiner" on

THE SEA-RAVEN (*Hemitripterus Americanus*).

an angler's rod—the unsavory bait is successful, and the small
fish proceed to their doom, as the great mouth opens and
takes them in, remorselessly as a scoop-net. The great number which they swallow often leads, however, to their own
destruction ; for, though this marine angler is not himself
desired for the table, fishermen often kill him for the sake of
the numerous small fish which are found in his stomach.
The apparatus with which he baits his victims is unique in

the piscatorial line. It is one of several long, movable spines, rising from the back of the head, arranged something on the plan of a hook and staple; it is very curious, and well worth the trouble of dissecting a specimen, to see the beautiful adaptation of Nature in this bony contrivance to help the ugly fellow to a dinner. Its *locale* is northerly, not being found

THE ANGLER (*Lophius piscatorius*).

farther south than the Capes of Delaware. It is frequently caught in New York waters in the early spring, but does not bear confinement well, and lives but a few days in the tank.

THE HAMMER-HEADED SHARK (*Zygæna malleus*).—The essential peculiarity of this member of the great family of sharks is the shape of its head, which plainly resembles the head of a hammer, while the rest of its body is similar to other sharks. The head is very much flattened, and from either side project arm-like protuberances, upon the very extremity of which the eyes are placed, so that, while one is directed to one point of the compass, the other necessarily looks in an exactly opposite direction. Thus this great marine cruiser (it often grows to the length of ten feet) may be said to always have a " watch set," both on the " starboard " and " larboard," to protect his precious body. By a peculiar arrangement of the eyelids, however, he can much more easily

THE HAMMER-HEADED SHARK (*Zygæna malleus*).

look up or down than in any other direction. It is viviparous. I have never succeeded in placing it in the aquarium alive, though I have made thorough and repeated efforts to do so. Whether they die from fright, or whether the eyes, which are in such an exposed position, and are organically connected with the brain, get injured by contact with the can or vessel in which they are transported, it is difficult to tell; but it is certain they do not " make an effort " to live after capture. They are sometimes found in Long Island Sound. Their range is from Brazil to Cape Cod, and except by accident they are not found north of the latter.

The SKATE or RAY (genus *Raia*) contains many varieties, such as the clear-nose, the spotted, the whip, the prickly-tailed or sting ray, and the smooth skate (*Raia lœvis*). Specimens of all these kinds of rays may frequently be seen hung up in the markets to attract attention, which they do from the ludicrous, serio-comic facial expression.   They are often

THE UNDER-SIDE OF SKATE.

taken in company with the cod, and are occasionally eaten by the buyers of cheap food, though their substance is chiefly gelatinous and the flavor disagreeable.   Some observers have imagined that the under-side of the head bears a resemblance to the human countenance, but we think no one would care to claim relationship with these nondescript-looking subjects. The whip-ray is so named from the use he makes of his very long, narrow tail, from five to six feet in length, which he slashes around in fine style when excited. The tail of the prickly or sting ray is beset with fine spines. The smooth skate or ray is distinguished by a nearly total absence of these spines. The eggs of the ray are of a curious shape— a sort of oblong, with four ribbon-like threads attached, one

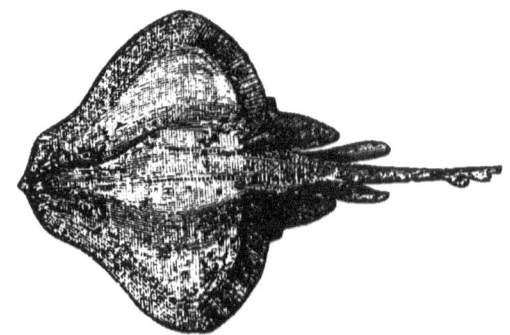

SMOOTH SKATE (*Raia lævis*).

at each corner; with these, aided by a self-generated mucus, the parent ray attaches the eggs to algæ, stones, or rocks. These empty egg-sacs, after the fish are hatched, are often picked up on the sea-shore by persons who never imagine what was their original use; in some localities they are known as sailors' purses. I have had skates' eggs hatched out in my aquarium, but the young lived only about two weeks.

HAIR-FINNED ARGYREIOSE (*Argyreiosus capillaris*).— This queer-looking little fish, faithfully delineated on the upper right-hand corner of the cover of this book, is sometimes caught in the harbor of New York. It is of a lustrous, silvery white, varying its attractions with opaline tints and sometimes brilliant iridescent colors. In size it rarely exceeds seven or eight inches, and it is oftener met with from

two to three. Its shape conforms to no recognized figure.
It is strange and even comical in appearance.   Its facial
angle descends with a slope of sixty degrees, until it arrives
at an angle of about fifteen degrees to its first dorsal.   It
looks, in fact, as if its forehead had been chopped off with a
broad-axe—a kind of distorted, beveled, bias, sloping face.
The expression of the living fish, too, is as singular as its
outside contour.   It seems to have a sort of "Uriah Heep"
meekness and humility, as if it were itself uncertain as to
the correctness of its make-up—as to whether it were *really
finished*. From its exceeding oddity it is a valuable acquisition to the aquarium.   Its wafer-like thinness adds to its
many peculiarities, and perhaps to its real lack of vitality :
its life in captivity seems to hang by a thread; a sudden
fright or surprise is sufficient to extinguish it.   It may be
found in considerable numbers in our waters in August; it
is best taken in a fike-net, on account of its extreme delicacy.

BALLOON-FISHES.—There are many varieties known by

BALLOON-FISH (*Chitomycterus geometricus*).

this name, distinguished as spotted, striped, spiny, etc. The colors also vary, but are often of a bright sea-green above, with olive-brown markings. They are usually more or less thoroughly furnished with prickles or spines over the body, mostly of a sharp, recurved, triangular, and compressed form. Their principal peculiarity consists in their ability to distend themselves like an inflated balloon ; this they will do when

BALLOON-FISH, OR SEA-PORCUPINE (*Diodon pilosus*).

taken from the water and rubbed with the hand, and they also assume the form to suit their own purposes when swimming. They are small, and may be taken with the hook in our harbor in July and August. This latitude is thought to be their northern limit. Their principal food appears to be soft young mollusca. There is some query among naturalists whether the many so-called varieties of balloon-fish may not prove to be simply different stages of growth of the same creature.

The carefully-drawn illustrations on the following pages, together with those contained in the foregoing chapters, will give the reader some idea of a few at least of those interesting salt-water animals which may be readily captured and successfully kept alive in large marine aquaria.

GREAT SUN-FISH (*Orthagoriscus mola*).

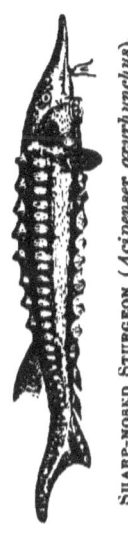

SHARP-NOSED STURGEON (*Acipenser oxyrhynchus*).

KING-FISH (*Umbrina nebulosa*).

TAUTOG OR BLACK-FISH (*Tautoga Americana*).

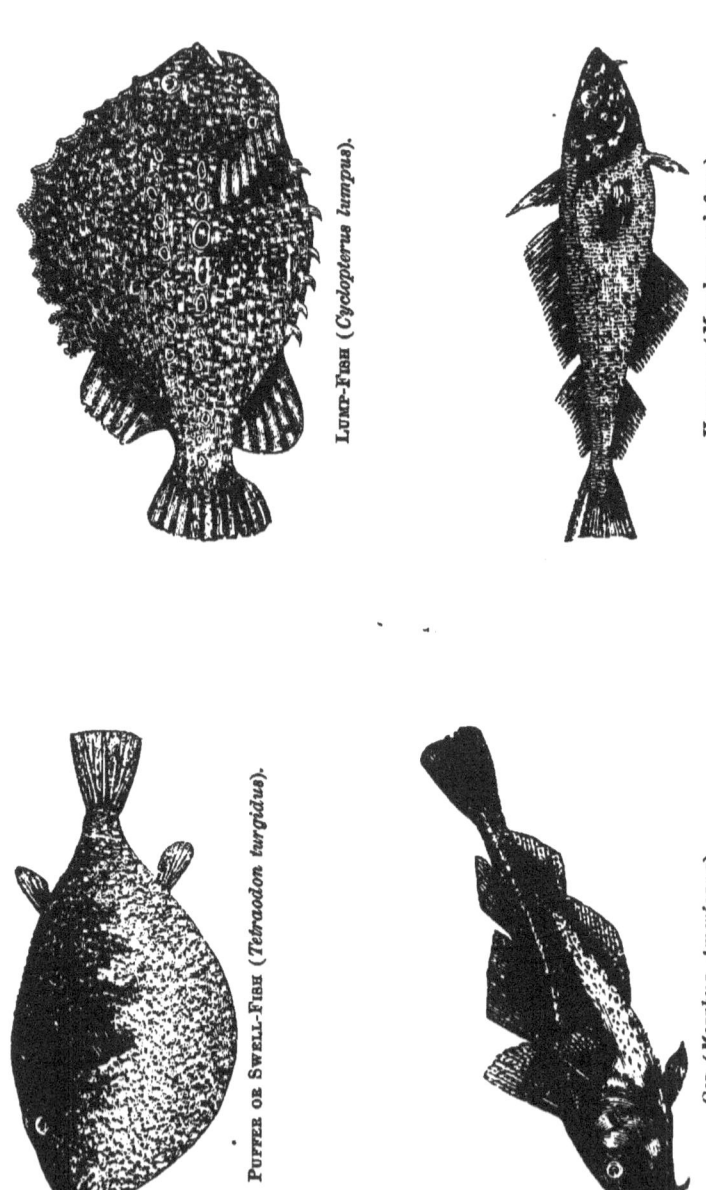

LUMP-FISH (Cyclopterus lumpus).

HADDOCK (Morrhua æglefinus).

PUFFER OR SWELL-FISH (Tetraodon turgidus).

COD (Morrhua Americana).

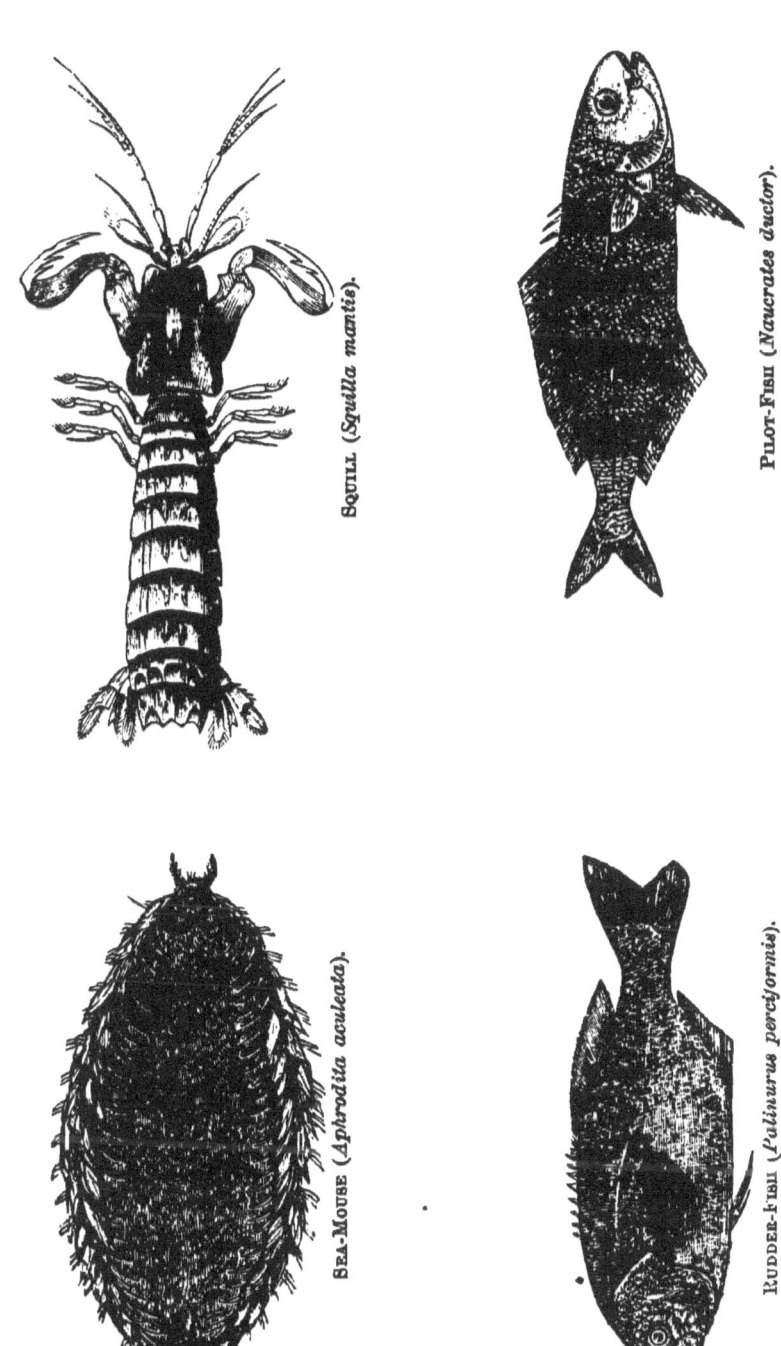

SQUILL (*Squilla mantis*).

PILOT-FISH (*Naucrates ductor*).

SEA-MOUSE (*Aphrodita aculeata*).

RUDDER-FISH (*Palinurus perciformis*).

TURTLES.—There are land and marine turtles, each exhibiting marked peculiarities; but those which we call marine come on the sandy shores to deposit their eggs, which are round, not oval. The hawksbill turtle (*Caretta imbricata*) is one of the most interesting and valuable; it is from this variety that the fine "tortoise-shell" combs and jewelry are made. Some mercantile houses —like Messrs. Tiffany & Co., of New York— sell annually thousands of dollars' worth of it. The upper shell of the turtle grows in plates, the pattern being varied in the different kinds; but in the hawksbill they are nearly heart-shaped, notched at the edges, the apexes pointing backward and overlapping each other. These shelly plates are reduced to any required shape by steaming and pressure; even small pieces are, by the heat and immense weight applied, so welded together as to appear homogeneous, and thus any degree of thickness may be produced. The scales taken from the under carapace of the turtle are used to make a kind of jewelry resembling amber. It is very scarce. The name of this turtle is derived from the hawk-like appearance of its horny beak.

HAWKSBILL OR SHELL TURTLE (*Caretta imbricata*).

GREEN TURTLE (*Chelonia viridis*).

Turtles have no teeth. Some live on algæ, others on

mollusks and radiates.   Their feet, particularly those of the hawksbill, form perfect oars.   They are sometimes met with hundreds of miles from land.   I have caught this variety in the Bermuda Islands; they are never found far north.

The green turtle (*Chelonia viridis*), so dear to the hearts of our gormands, is well known for its edible qualities; in commerce its shell is used principally for button-making.   It is often taken when sleeping on the surface of the water, and great numbers are captured when they come on land for purposes of incubation.   When hatched the instinct of the young leads them straight to the sea.

CHAPTER X.

EVERY one has heard of one class of barnacles, those which attach themselves to ships and thus hinder their progress through the water; but few landsmen have ever had opportunities of learning the habits of this animal from personal observation. Indeed, it is not rare to meet with persons who have no clear idea of whether they are animals or a mere aggregation of shells, as they may have seen them exhibited in some collection of curiosities. And this fact recalls another, which results from the precedence and importance given to mere literary education, apart from the practical and scientific, namely: that there are yet many otherwise intelligent

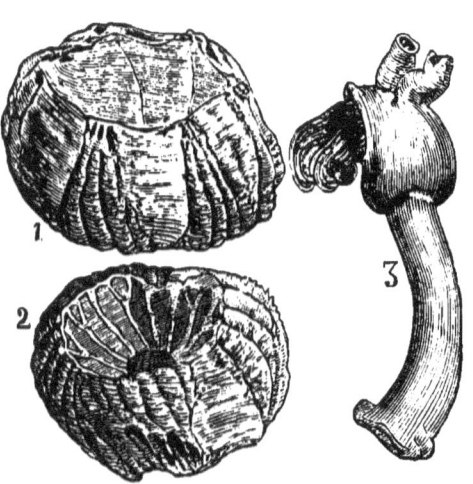

SPECIMENS OF THE GROUP OF CIRRIPEDS.—1 and 2. *Coronula diadema* (sessile). 3. *Conchoderma aurita* (pedunculated).

people who think that shells are a production of the ocean, without once imagining that they are the mere cast-off covering of some variety of mollusca. The so-called "art of conchology" proves this: the art of arranging, classifying, and indexing *mere shells*, as still taught in some schools, has naturally fostered this unscientific error.

The scientific name of the barnacles, *Cirrhopoda* or *Cirripedia*, means curled or hair-like feet (*cirrus*, a curl or lock of hair) ; and if they had occasion for feet, it is the *cirri* with which they are furnished that they would use for pedestrian purposes.

The name of *Anatifera lœvis* was originally given to the ship-barnacle out of regard for an old legend which attributed the production of the barnacle to the bernical or solan goose, a theory so preposterous that it seems incredible how it should have been believed for hundreds of years, and that witnesses presumed to be trustworthy should have minutely described, and even illustrated by drawings, the whole process of the evolution of the sea-fowl from the barnacles attached to trunks of trees, which had first been carried out to sea and then drifted ashore on the coast of Scotland ; for it is in that land of " second-sight " that the tradition originated. The historian Hector Boece in 1490 declared that he *saw* the larvæ of the barnacle "*formed into perfect foules ;* " and this was indorsed by the parson of the parish. One hundred and fifty years later (1636) Master Gerard gives another circumstantial account of the same mysterious evolution ; and it would probably not be difficult to-day to find in the more secluded portions of the coast old fishermen who would aver that they " had heard the cry of the young goose out of the barnacles ! "

In a strictly scientific classification, the *Cirrhopoda* should be placed between the annelids and the crustacea, while judged by the shell alone they would be ranked with the mollusks. The shell is of the multivalve species, though bivalve in appearance, each side being composed of two plates connected by a long central plate which joins the two halves at the back. Some varieties are *sessile*, that is, seated or attached closely to the object to which they ad-

7

here ; but they are mostly *pedunculated*, that is, connected by a cylindrical, flexible, annular stalk, of a fleshy nature, varying in color from a bright orange to a purplish red. This formation will be better understood perhaps if we describe the transformation which takes place in the young barnacle before it becomes fixed in its adult and sedentary life. These changes are almost as wonderful as those imaginary ones described by the old chroniclers.

Like some other marine existences of a low order, the young of the barnacle is, in its freedom of movement and some other particulars, a more highly-developed being than its parent. In its earliest form it is furnished with a broad carapace or shelly covering, two pairs of antennæ, three pairs of legs, which are branched, jointed, and supplied with bristles, and a forked tail; it has also a single eye, cyclops-fashion, in the centre of its anterior portion. It moults, casting off its shelly skin three times before attaining to adult size ; and at each moult great changes in form may be observed. At the third the carapace has changed into a bivalve shell; the head with its antennæ is greatly enlarged; the one eye has disappeared, and two much larger ones take its place. Now comes the most curious part of its history. Just as it appears to be fitted out and equipped for enjoying a free existence, its fate compels it to surrender all these advantages, and it sets out to find a suitable place to complete its last permanent metamorphosis. It finds a floating piece of timber, meets with a ship becalmed, encounters a whale lying at ease, or perhaps a turtle whose back seems fitted for the accommodation of parasites : on some hard, firm substance it affixes itself.

One would think that in the ever-moving waters it would be no easy thing for one animal to attach itself to another, or to fix itself upon a rock, or any substance continually subject to the buffeting of the waves. The secret of it lies in the capacity of the creature to produce that *wonderful marine cement*, insoluble in water, such as some of the annelids (described in a former chapter) use, and which enables the young barnacle to attach itself to whatever object it selects.   This is a kind of organic gum secreted in the anterior antennæ, which the animal pours out when it seeks to attach itself, and which hardens immediately; and with this it chains itself in perpetuity, a self-condemned galley-slave for the remainder of its life.   But further changes occur: being now adherent by the fore-part of the head, it throws off its bivalve shell and discards its eyes, which henceforth would be of no use to it; the head lengthens out; the new shells arc five-plated; the legs are transformed into *cirri* or curling tendrils, which now operate in the reverse direction, and are used solely for the purpose of creating currents to draw infusorial food toward the mouth, which is henceforth its sole employment.   Surely these changes, which science has verified, are as wonderful as any which ignorance could invent, aided by superstition and self-indulgence; for it so happens that the old legend of the barnacle-goose being hatched from barnacles was seized upon by the Romish clergy as a pretext for eating these fowls on fast-days—the reason being given that "they were not produced from flesh of any kind, and therefore might be eaten as fish."

Barnacles may be of any size, from a pin's head to two or three inches across, and possibly even larger.   Their *cirri* present a very beautiful appearance when the creature waves them forth.

The most common and beautiful are the acorn-barnacles (genus *Balanus*, a Greek and Latin word meaning acorn), so called from their resemblance to the seed of the oak. They are often found on the rocks of the sea-shore, and sometimes reveal themselves in a very uncomfortable manner to the bather who attempts to walk barefoot over a colony of them upon a rocky shore, or to clamber over their sharp-edged domiciles by the aid of his hands.   Those white coni-

cal objects, which look so innocent, are almost as sharp as
razors; and if you have far to travel over them, your feet
will be cross-hatched like an engraver's plate. This variety
are very apt to attach themselves to the backs of shell-fish,
crabs, and even whales. Their shells are composed of six
instead of five plates; their general form being nearly circular. These creatures will burrow two or three inches through
the skin into the blubber of a whale. Sometimes they are
found in tide-pools in the rocks, and when the tide is low
they present a very unattractive
appearance, remaining closed up
without signs of life, looking
like a mere incrustation upon
the rock; but when the tide returns they very promptly regain their energy, and soon from
every tiny shell a graceful *feathery hand* is stretched forth like a
little fishing-net, elegantly waving in the right direction to propel animalcula to their digestive
apparatus.

GOOSE-BARNACLES ATTACHED TO A
BOTLLE.

Almost every kind of marine
creature has its own peculiar
mode of foraging. Some throw
out fishing-lines like some forms
of medusæ; some rush after their prey and seize it by sudden violence; others lie in ambush, and try to conceal their
presence, like the octopus; and thus the barnacle throws out
its casting-net, drawing it in and contracting it at intervals.
This net is formed of a group of fine, delicate tendrils, called
the *cirri*, each double at the extremity; they are jointed so
as to make an exquisitely-shaped curve, and at each joint are
long, stiff hairs. As these *cirri* are quite numerous, and the
hairs upon them stand out firmly when in action in a trans-

verse position to the *cirri*, the whole apparatus forms a complete network, so fine that the most diminutive animalcula
would be held within its toils. The *cirri* themselves are of
a horny nature, though rendered thoroughly flexible by
ample jointing.

I have kept barnacles for many years. During their captivity they do not seem to have suffered from any cause, as
the fact of their growth proves, some of them having increased to double their original size. I do not know the
limitations of their growth, but certainly it is quite rapid
even in confinement. Some shells in my possession, which
I use for *flower-vases*, stand fully four and a half inches in
height; and my enthusiastic aquarian friend, Mr. Roberts,
assures me that he has seen them used very commonly at
Wood's Hole, Massachusetts, for inkstands. I have always
noticed, however, that the small ones are the most active, the
larger ones remaining closed and quiet the greater part of the
time. But sometimes, when they have been entirely passive for
a considerable time, I can coax them to disclose their feathery
hands by dropping a little clam or oyster juice into the water
near them. Then, where a moment before all was perfect
calm, the surrounding water will be full of fairy-like fingers,
beckoning their food toward them; and so rapidly is this
peculiar motion made, that a very perceptible current in the
water is produced, which draws every little floating particle
toward their open mouths.

Three or four times, during many years of the closest observation of these interesting animals, I have noticed a curious demonstration going on between the members of the
group—a queer kind of hand-shaking, or some occult com-
munication, which would no doubt be exceedingly interesting if they would explain to us its meaning, but which at
present is not perfectly understood by naturalists. In this
sudden awakening of the social instinct the barnacle does not
use its regular net-like hand, but puts forth a single long

tentacle, reaching over and among a dozen perhaps of its neighbors, extending a distance of some inches. Sometimes it penetrates with this into the openings of the other shells, as if it would inform itself as to their continued existence or condition of health; but, having finished its inspection, it quickly retracts and hides within its own shell.

The barnacle sheds its coat at intervals like the crab; but much more frequently does it discard the thin, transparent epidermis which covers the beautifully-barbed fingers. This it throws off as neatly and completely as one could remove a nicely-fitting glove from the hand. These cast-off " barnacle-gloves," with their minute hair-like barbs, form an interesting object for examination under the microscope.

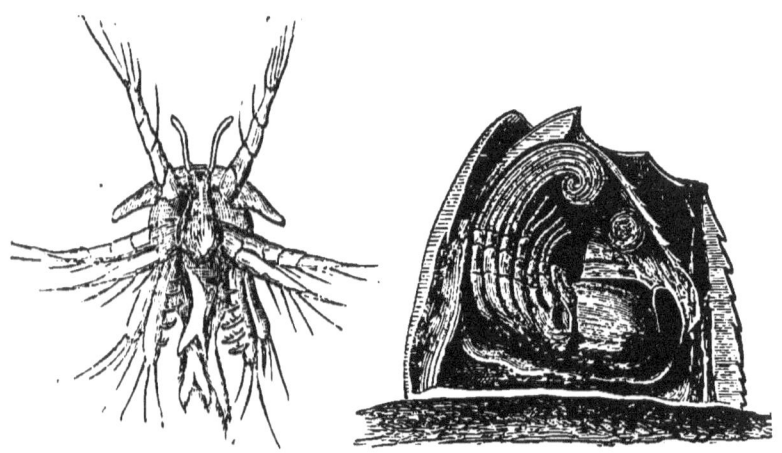

YOUNG OR FREE-SWIMMING CIRRHOPOD.   ADULT CIRRHOPOD, WITH ONE HALF ITS SHELL REMOVED, SHOWING ANIMAL AT REST.

CHAPTER XI.

ONE of the most common objects to be met with at New-port, Nahant, or almost any point along the Massachusetts coast, are the so-called "star-fishes," though scientifically speaking they are no more fishes than is a rabbit or a bird;

STAR-FISH ON A ROCK.

yet, for convenience and to save circumlocution, we may adopt the popular name in speaking of them. At low tide these curious "stars" may be seen by thousands, sometimes clinging to the rocks, sometimes on the gravelly bottom, or perhaps attached to the sea-weeds.

Some of those which I have had were colored a dark, rich, velvety brown ; others are of a reddish cast, and some again of a chocolate hue. But, however tinted, to the naturalist who looks beyond the *complexion* they are always attractive, and in their structure and habits favorite subjects for examination. The late Prof. Agassiz seemed particularly fond of watching and investigating the proceedings of these animals, and spent, I may safely say, years of labor working up his exceedingly careful and valuable account of the great sub-kingdom of the *radiates*, to which the stars belong.[1] In the large folio edition of his "Contributions" may be found numerous and most elaborate drawings showing the internal organism of these and other marine animals. I believe he was just finishing the work referred to at the time of his lamented death.

The fact that an animal is common by no means implies that it or its habits are well known ; the fact is often the very reverse. If an object is brought from five thousand miles over sea, no matter how intrinsically worthless, it can generally secure admirers ; while some of the most wonderful productions of Nature often lie unnoticed at our very doors. Until lately this might be truly affirmed of the common "five-fingered jack," as seamen call the star-fish. One reason for our ignorance of this class of marine animals is found in the fact that, previous to the general introduction of aquaria, naturalists could rarely get anything but the skeletons of these creatures to examine—a very inadequate means of learning the habits of the living animal.

If we understand the principle upon which the five-rayed star-fish is constructed, we shall readily perceive that all the other varieties are simply modifications of this. The first thing to be learned about this animal is that it has *two distinct sides*—an upper, slightly convex, and an under or oral

[1] *See* "Contributions to the Natural History of the United States," by Louis Agassiz, vol. iv

side. The upper is rough and tuberculous; the under is soft, and contains all the vital and locomotory organs. It is divided by five furrows, all leading to the mouth, which of

STAR-FISH, SHOWING PROCESS OF REPRODUCING ITS FOUR LOST ARMS.

course connects with the digestive cavity; in this it some-what resembles the actiniæ, but it differs wholly from them in the relative powers of expansion and contraction, as well as locomotion.

The rays of the star-fish are usually on the same plane, but the animal has the power of raising them so as to pro-gress over obstructions which may be in its way. It can walk on a level or ascend elevations with apparently the same ease. But it is not the rays themselves which perform the work of locomotion. Its organs of motion we shall presently de-scribe; but, in the first place, let us see how this star-fish looks when first taken from the water. Almost immediately the soft under parts seem to shrink away, and nothing substantial remains but the upper surface; this is perforated with pores, through which the water enters to all parts of the body by specific channels; it is a sort of tidal flow of water—alternate absorption and rejection—which occasions the motions of the rays. Upon the surface of the body, near but not precisely at the centre, is a small opening through which water is admitted to a strong, elastic tube, which tube is encircled by a series of rings. Now let us turn our star over, and we shall see that this tube opens into a ring about the mouth, while similar tubes stretch the length and breadth of the arms. From the cross-tubes fibres extend, terminating in disks, which are also continued entirely through the animal into the furrows before mentioned; these are the true organs of locomotion, and they are called *ambulacra*. The mode of progression is something like that of a ship dragging its anchor; thus a portion of these ambulacra are made fast by suction, while the remainder of the body is drawn forward, when the first are relaxed, and then the process is repeated. The pace is slow, and the description would make it appear clumsy; but the reverse is true, as the motion has a certain degree of ease and grace, approaching to elegance, strictly following all the indentations of the ground. That when walking they *perceive* obstructions is evident, for they immediately prepare to surmount them. How much they can really *see* with their five eyes is doubtful, for their organization is exceedingly low; but as they are placed one at the end of each ray, in a sort of "round-robin" fashion, no particular ray can *put on airs* and pretend to take precedence of the others; and our star may travel with either of the five rays in advance. These eyes are the bright-red spots near the tip of each arm or ray.

In addition to the digestive and locomotory systems, the stars have a heart, situated near the opening on the back, which is supplied with a set of blood-vessels. There are also a respiratory apparatus and a nervous system; but the latter must be of a very low order, judging from the capacity this animal enjoys of enduring vivisection apparently without pain.

One of the interesting traits about this lowly-organized creature is the care it bestows upon, and the evident affection it feels for, its eggs. These are contained in pouches situated at the broad base of the rays; and, when emitted through an opening there provided, far from abandoning them, as many fishes do their young, these stars gather them together, bending their arms downward, and at the same time arching the ·central part of their body, thus forming a sort of protection for them—absolutely "brooding" them, chicken-fashion. If the eggs are accidentally scattered, they will take great pains to collect them again; the experiment has been tried over and over again in the aquarium, and they will travel the whole distance of the tank till they find and recover them.

If in the course of their search, say in crossing the tank, they encounter a piece of rock, they do not go round it; but, elevating one ray, they gradually draw the others after it, and by slow degrees rise over and make the opposite downward movements with as little apparent effort as if walking upon a plane. When they take the whim of ascending the smooth glass sides of the aquarium, it is all the same; but in descending they sometimes allow themselves to drop, by way of variety.

Their mode of reproduction is not limited to eggs. They have the strange capacity and frequent habit of detaching one or more of their rays, when each of these cast-off members becomes in time a perfect star. I have seen this operation performed many times, almost incredible as the statement seems to those unfamiliar with the vagaries of the zoöphytes and radiates. For instance, an arm or ray would perhaps be accidentally broken off close up to its point of junction with the central portion of the body. The animal, instead of appearing to be disturbed or annoyed, as it would be at the loss of its eggs, appears to mind the disappearance no more than if it were a cast-off garment, and goes about as happy with its remaining rays as if the whole had remained intact. Perhaps if we could replace a lost arm as easily as our star, we should be nearly as indifferent to such a loss; for what do we see next? Only a little protuberance where the lost arm was separated. But look again in a week, and we shall see some little suckers or ambulacra projecting; the parts by degrees enlarge, and at the end of a few weeks a somewhat smaller but apparently quite perfect arm takes the place of the lost member! Its spines, water-tubes, tentacles, pedicellariæ, etc., are all in perfect working order, and its normal functions are fulfilled with all the precision of the elder rays. It is not, however, *quite* equal to the original; besides being smaller, it is of a more delicate texture, and its color of a lighter shade. It is very interesting to watch this extraordinary effort of Nature in the development of the new member. The last one I had in my collection was just fifteen weeks in producing a new, full-grown arm.

Star-fishes found upon the shore often appear to be quite dead when they are really alive; they are the opossums of the sea. Take up one of these fellows who is lying perfectly still, and put him into fresh sea-water, and he will very likely soon be traveling about as well as ever. However, as the dead and living, when left stranded by the tide, present so nearly the same appearance, it may be well to have some test by which to make sure of their true condition. There are two modes of ascertaining this with a reasonable degree of certainty. If, on taking up a star-fish, he hangs loose and limp, he *is* dead; but, however dead he may look, if on touching it there is a firmness and consistency in the substance, he is only "playing 'possum," and will revive in the water.

The other mode of trial is to lay our starry friend on his back, when if he is alive you will soon see a number of semi-transparent globular objects beginning to move, reaching this way and that, as if feeling for something; these are the loco-motory organs or ambulacra, seeking to regain their normal position. If there is no movement of these, you may con-clude that he is an extinguished star.

The main stomach belonging to this animal appears too small to contain the quantity of food which he often devours, and would indeed prove so were it not supplemented by sev-eral additional stomachs which occupy the several rays through nearly their whole length. These supplementary stomachs are called *cœca*, and explain in some measure the voracity of these animals.

The small whelk, *Buccinum obsoletum*, is a favorite dish with the star-fishes; they eat millions of them. Almost every one knows this little mollusk by sight; it is a small, ·black-shelled, snail-like object, usually about five-eighths of an inch long, and may be seen on almost every stretch of shore on our coasts. I have seen a star-fish with no less than six of these in its mouth at once.

In attacking small mollusks, star-fishes often envelop the victim completely with their arms till they get him snugly fixed in the mouth, when they relax their rays and proceed to suck out the fleshy substance at their leisure. This is all quite comprehensible; but not so easily understood is the mode by which they succeed in destroying large bivalves like oysters; yet they do this to such an extent as to prove disastrous enemies of the oystermen. Clams and mussels also suffer literally " at their hands." The *fact* was long recog-nized before the mode of attack was comprehended. Within the jurisdiction of the Admiralty Court of England there was, and may be still, an old law affixing a severe penalty upon those who " do not tread under their feet, or throw upon the shore, a fish called a five-finger, resembling a spurrowel, because that the fish gets into the oyster when they gape open and suck them out." It would have been well if our own oystermen had observed this sensible law. Instead of this, many of them, and also fishermen, have been in the habit, when star-fish were brought up by nets, rakes, or dredges, of tying them up in bundles, and drawing the cord tight enough to cut into the whole pile ; and, supposing that thus they have certainly made an end of their worthless lives, they throw them overboard into the water again, not realizing that each of the pieces into which they were divided would in time become a perfect star-fish, thereby increasing their own and other poor fishermen's trouble—five times possibly. Where they had at first say one thousand enemies for their oyster-beds, they, through ignorance, have increased them to five thousand !

The manner in which the star-fish attacks the oyster is unique in its way. Instead of inserting a ray and thus drawing the oyster out, as was formerly believed, a closer observation has evolved the fact that the star has a trick of partially protruding or pouting out its own stomach, and that it actually thrusts or insinuates this between the edges of the bivalve shell, and by the power of suction destroys the oyster, consuming it utterly in spite of its strongly-protected condition. The query has always been, why the oyster did not close its shell and entrap the star ? One reason may be, that the oyster is not sufficiently sensitive to resist the first approach of its enemy, and the star is, as we know, very insensible to pain ; so that a slight vantage gained at the outset by the assailant would go far toward accounting for the easy victory which it gains over the sluggish bivalve. If any one has ever watched the careful way in which a star advances and softly crawls over his prospective dinner, the mystery would not appear insoluble, even though the victim was protected by an apparently invulnerable calcareous shell.

CHAPTER XII

ECHINOIDS AND SEA-CUCUMBE

The sea-egg or sea-urchin may be considered the representative animal of the *Echinodermata*, of which star-fish and others are less pronounced members. The word *echinoid* means possessing an exterior *resembling a hedge-hog*. It is from structural resemblances, and not from shape or superficial likeness, that marine animals are classed. Though

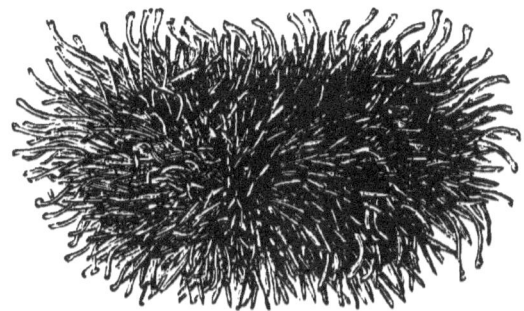

SEA-EGG OR SEA-URCHIN (*Toxopneustes drobachiensis*).

sometimes the long, hard names have a tendency to repel the reader, yet when understood they are usually very expressive; indeed, they often suggest a whole history, sometimes a romance, or the lifetime devotion of friends. A few instances occur in which scientific names have been given showing the meanness of spite or envy; even the great Linnæus condescended to this when he named a valueless, offensive weed after Buffon. Could the origin of all the scientific names now used in the various branches of natural science be written out, it would form a most entertaining book.

But to return to our sea-urchin. This curious animal is of a spherical form, and covered all over with long, beautifully-shaped spines. Its shell or covering resembles a limestone network, but is in fact formed of separate plates, so neatly joined that the sutures cannot be perceived from the outside; but, on examining the skeleton from the inside, the points of connection may be traced. As the urchin does not cast its shell, the question naturally arises, "How can the creature grow, enveloped as it is in this inflexible substance?" The answer reveals a wonderful contrivance which admirably meets this difficulty. Over the shell we find a very delicate membrane, vital, as its function proves; this is insinuated between the jointed plate-armor of the carapace, and steadily deposits there a secretion of calcareous matter, so that each separate plate is simultaneously increased in size; and thus the animal is enabled to expand until the adult size is attained, when the secretion ceases. As usually found, dead on the shore, the urchin is devoid of spines, and presents something the appearance of

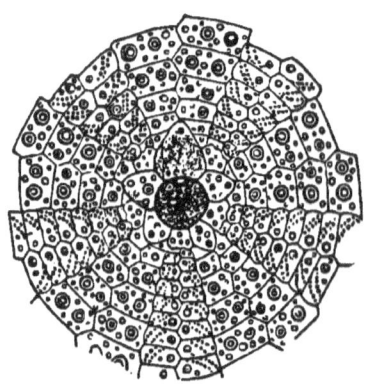

SEA-URCHIN, OR SEA-EGG.—Top view, spines removed.

a melon, the surface being marked with ten zones or divisions, five being larger than the alternating moiety.

The urchin's relationship to the star-fish may be illustrated by supposing that we bring all the five points of the star together, filling up the interstices with a similar substance : we have then a complete urchin, minus the spines. Or, take the peel whole off of an orange, divide it into fifths, and bring the points up together, sticking needles in to simulate the spines, and we have an urchin, at least in shape.

Like almost every known animal, this echinoid is much handsomer when alive and sailing about.   The color is usually reddish brown or black.   I have secured specimens of them in Bermuda with long, tapering spines a foot in length, and standing boldly out "like the quills upon a fretful porcupine ;" the body or ball part not being larger than a hen's-egg.   Indeed, when these spines drop off and leave the shelly covering exposed, they very much resemble an egg, which accounts for their secondary name of sea-egg.

When alive they are very shy, concealing themselves in holes and crevices of the rocks.   They even go beyond this, and attempt complete seclusion from observation by covering themselves with bits of sea-weed, sand, or anything they can get hold of for the purpose, by means of their tentacles and pedicellariæ, so that an inexperienced person would scarcely recognize them.

The upper portion of this circular animal—what we might call its "north pole," where the ribbed zones unite like the degrees of longitude in a map of the world—is called the "dorsal area."   On the under-side, or "south pole," the mouth is situated.   This circular orifice is furnished with five strong, hard, sharp teeth, resembling flat conical wedges, with numberless perforations, through which pass long, slender, delicate tentacles, terminating in suckers ; these are scattered all over the surface of the investing membrane, between the spines.   In addition to these, the urchin is provided with another set of weapons called *pedicellariæ*, each consisting of a long stem on the end of which is a kind of knob, capable of opening like a miniature ·trident or three-tined fork, the tines being concentrically arranged; when closed, these little prongs fit into each other like nippers. Curiously enough, these pedicellariæ were for long mistaken by naturalists for parasites upon the body of the urchin.

These animals are voracious vegetarians, eating off large fronds of the sea-lettuce and other plants, and cleaning a tank of every vestige of vegetation in a very short time. Their motion in swimming is slow, and when walking on the side of a glass tank, which they do with perfect ease on their long, slender legs (which are terminated by cup-shaped disks, constructed on the same principle as a surgeon's cupping-instrument), and aided by the spines, they are certainly an attractive sight, especially when all the spines and numerous pedicellariæ are fully distended. To the cursory observer they look no more capable of ascending a smooth surface like glass than a chestnut-bur does of walking up the side of a house.

Echinarachnius, the sand-dollar or sand-cake, is a modest relation of the urchin. Instead of being spherical, it is flat,

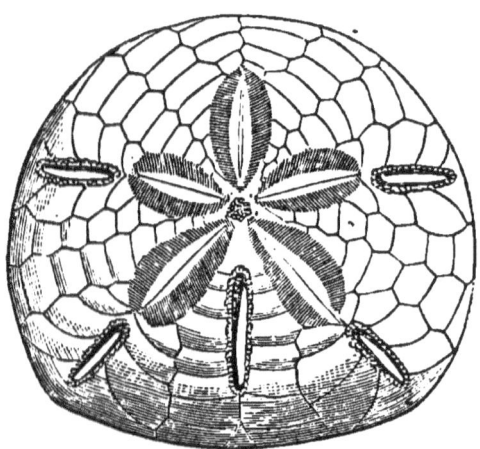

KEYHOLE-URCHIN (*Mellita-quinquefora*, Agassiz).

and really looks when at rest something like a circular cake of sand, about the size of the old silver dollar—perhaps a *little larger* than the new issue ! It is a reddish-brown color, and covered all over with short, hair-like points, resembling a piece of light velvet when viewed across its convex surface. These expand near the dorsal centre or summit into a rosette-like disk; when the fine, spiny projections are removed, the ribbed divisions of the surface are found to be identical in general direction with those of the spherical brotherhood. Mrs. Elizabeth C. Agassiz, in her "Sea-side Studies," says that this flat variety of the echinoids belongs to the family known as the shield-like sea-urchin (clypeastroids). When speaking of the *teeth* of the *Echinodermata*, we should have

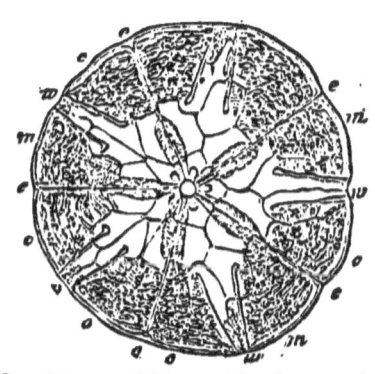

CAKE-URCHIN (*Echinarachinus*) : *o*, mouth; *e e*, ambulacra ; *c m*, ambulacral ramifications; *w w*, interambulacra (Agassiz).

mentioned that they have a concentric motion—all converging toward the centre, never moving up and down as with vertebrate animals. These sand-cakes are found all along the Massachusetts shore, but do not extend to Long Island Sound or New York Bay.

Ophiurans are related to the star-fish, and their rays or arms are five in number, like the common variety, *Uraster rubens*, but totally unlike in general appearance. In the common five-fingered stars, the broad bases of the rays form the central body of the animal; but with the ophiurans the arms are long and slender, nearly cylindrical in form, being slightly narrowed toward the extremities; they appear to be attached to, rather than to be growing out of, the small disk-like body. The spinous

SERPENT, OR BRITTLE STAR-FISH (*Ophiopolis*).

projections, which fringe the edges of these long arms, suggest the idea of five centipedes placed at regular intervals around the disk, and wriggling about with the intention of twisting themselves off if they can; and indeed they often do succeed, or rather the creature throws them off if much frightened, or, to escape capture, it will sacrifice one to save the rest. Like some others of this class, it has "suicidal tendencies." It is a very free swimmer, its centipedal arms forming excellent oars.

Their mode of progression on the shore or over rocks is like that of the common star, by putting forward one arm first, and then drawing the others after it. Their tentacles do not terminate with a sucking-disk, as in the star-fishes and urchins, but are covered with small tubercles. Neither have they teeth like the sea-urchin, but as a substitute they are provided with a bony plate at that end of the ambulacra connecting with the oral orifice. The ovaries are situated in the arms near their junction with the disk; small slits or openings may be observed leading to these pouches, and through them the eggs make their egress.

I have never succeeded well with the ophiurans in confinement. When taken in the net or dredge, they will throw off one, two, or even three, of their rays, so that it is almost impossible to secure perfect specimens to be deposited in the glass homes we should be so happy to provide for them; and the consequence is, that these self-mutilated animals rarely survive long in the tank. However it may be in their native habitats, they do not readily reproduce their rejected limbs in captivity.

There is one curious variety of the ophiuran family, found like the above in Massachusetts Bay, which has a habit of standing on the points of its rays—"tiptoe," as it were. In this position, with its tentacles hanging down, it presents the appearance of a basket; and hence "basket-fish" is its popular name.

The *Holothuria* or sea-cucumber is a curious cylindrical animal, varying in length from an inch to between three and four feet. There are several varieties of sea-cucumbers to be found on our shores, but the edible kind are most nu-

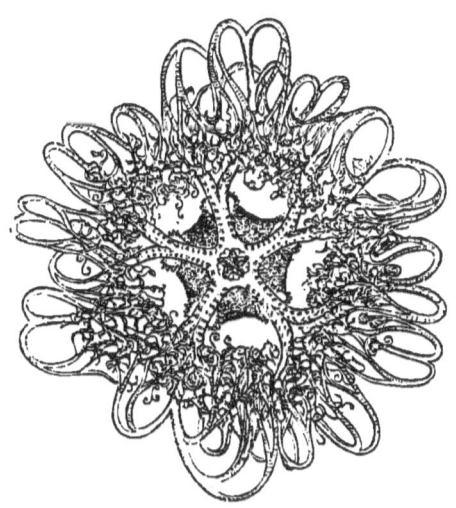

BASKET-FISH (*Astrophyton Agassizii*).

merous in the China seas. They inhabit deep water, but when found near the shore are usually partly imbedded in the muddy bottoms. Their outside covering is a tough, leathery skin, plentifully studded with short, hairy spines. The mouth, a circular opening at one end, is furnished with a wreath of beautiful plume-like appendages, which are extended at will for the purpose of grasping food and conveying it to the mouth; but, the food being brought within reach, only one of these tentacula is occupied in actually introducing it within the orifice, while the others remain passive, and appear to be waiting their turn to do the same service. Mrs. Agassiz has likened this group of tentacula in the sea-cucumber to some of the delicate sea-weeds, for their fineness of structure and the richness of their colors.

 · This animal has the ambulacral suckers like the star-fishes and sea-urchins; also the madreporic body, which is situated near the mouth, between two of the ambulacra, and opposite the fifth or odd one. Some of the species are also furnished with small hooks or fangs with which they can attach

SEA-CUCUMBER (*Holothuria lutea*).

themselves to the fronds of algæ, but their hold is not very strong. It has, like some other members of the *Echinodermata*, the capacity of fission, and will sometimes begin to contract in the centre, and finally divide itself, two perfect *Holothuria* developing from the one. Still more remarkable is the capacity of the creature to empty itself of nearly all its internal organs, and after an inconsiderable period to reproduce them and live on as comfortably as ever.

In the Bermuda Islands these animals grow to a very large size, and are mostly of a brown or nearly black color. They are so numerous there as to prove rather a nuisance to pedestrians on some portions of the shore. I have taken them weighing as much as twelve pounds and nearly two feet in length. They are sometimes eaten by the natives, but to my fancy a good appetite would be required to really enjoy the dish. Some varieties secrete an acrid and corrosive fluid, which lubricates the whole body; and these cannot be handled with impunity, as the contact produces a very disagreeable burning and tingling sensation. If it were not for their dark color, the name of "sea-cucumber" would be very much more appropriate to them, as at a little distance they bear considerable resemblance to that vegetable.

CHAPTER XIII.

HYDROIDS—MEDUSÆ—JELLY-FISH—PHYSALIA OR PORTUGUESE MAN-OF-WAR.

THE class of marine existences known in science as hydroids or animals which are developed from them, embraces a very large class of objects and includes many species which are little known even in their perfect and adult state; while their transformations and alternations in the course of repro-

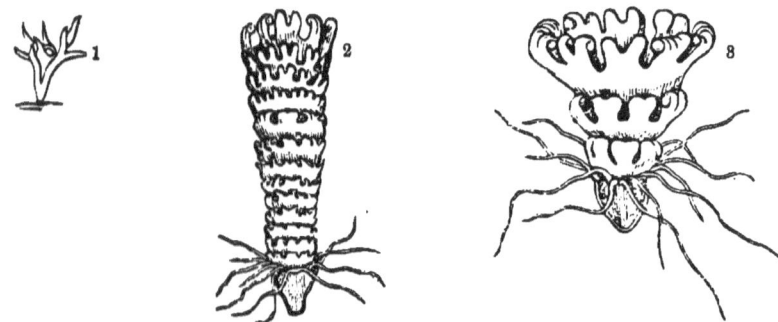

1. Early stage of Jelly-Fish (*Aurelia*). 2. Strobila, more advanced stage. 3. Strobila ready to be detached and form the adult (*Ephyra*, Agassiz).

duction and growth are so anomalous as to have long baffled the investigations of naturalists. Their nature is now better known, but we suspect there is much yet to be learned in regard to them.

The hydroids are *compound animals* which *produce individuals;* and these individuals in their turn produce com-

pound progeny.   Some of them look like shrubs, young
trees, or bushes, and only minute investigation proves them
to possess vitality, and to be in fact communities of individuals arranged in a plant-like form; from these may be
hatched a single jelly-fish or medusa, and from the medusa
will eventually be produced a group of hydroids.

The hydroids belong to the division *Acalephæ*, and are
as a general thing semi-transparent, fragile, and often very
graceful objects, yet not always so harmless as they look.   In

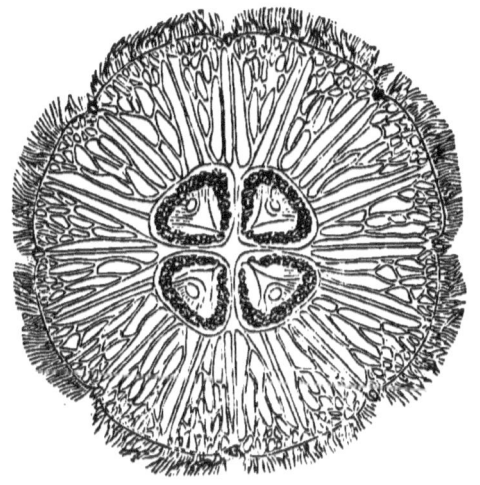

ADULT SUN-FISH, OR JELLY-FISH (*Aurelia*, Agassiz).

this division are included the large families of the campa-
nularians, sertularians, tubularians, and others.   The variety
of their forms is almost infinite.   Some float through the
seas like a branch covered with bell-shaped flowers; others
appear like groups of disks or cups; some are vase-shaped,
some oval; some are iridescent spheres; some form gar-
lands, or trail along like endless fringes; while possibly the
most common and best known form is the umbrella shape,
of which the beautiful haired medusa is the finest representative.   8

When the hydroid is in the compound state, the plant-like community is enveloped with a thin covering of a horny nature, which also covers the bell-like cellules in which the progeny are developed. When the latter arrive at a sufficient degree of maturity, this covering bursts and allows the escape of the young medusæ. These productive cellules are always developed in the angles formed by the junction of the branches with the stem of these shrub-like animals.

The *oceania*, a very pretty form of jelly-fish produced from hydroids, begins its free life as an almost invisible, shapeless little disk. Two minute tentacles may be perceived on close inspection, and two fine tubes forming part of its digestive system. Soon the shape becomes more distinctly spherical, and a feeding-tube depends from it; next numerous tentacles are added, "eye-specks" adorn its circumference, and, finally, some thirty-six long, fringe-like tentacles are furnished and Miss Oceania sails forth as unlike to her parent as a parasol is to a tree, and will in her turn produce descendants utterly unlike herself, but with the characteristic features of their hydroid ancestors.

Another very beautiful variety is the "ruined tower," *Turris neglecta*. This animal is dome-shaped, with a crest resembling a bell of red glass, which bears four rays in the form of a cross; from the edge of the bell depends a beautiful white fringe. It is a most attractive object when seen fully expanded.

The hydro-medusæ are so slight in their structure that they cannot be submerged, though they are often thrown upon the shore, and in that manner suffer shipwreck. When taken intact they will at first weigh surprisingly heavy for such transparent-looking objects, yet the weight consists almost wholly of nearly pure sea-water; but when the animal is stranded and dies, this water all escapes, and nothing is left upon the sand but a filmy, gelatinous skin, scarcely observable, or looking like flakes of dried varnish in the sun.

Some of them when distended with water will weigh ten or twelve pounds; others are so small that a few ounces of water will contain thousands. Some of these animals have a much denser fibrous organization than others ; some are so extremely delicate that one would feel no substance if moving the hand through water in which they were sailing and

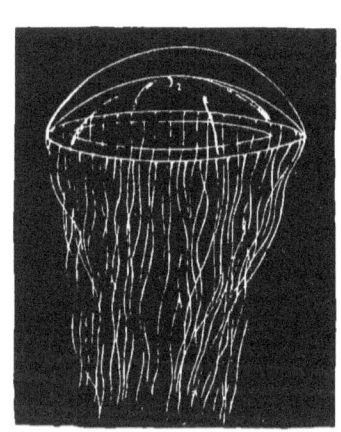

ADULT OCEANIA (Agassiz).

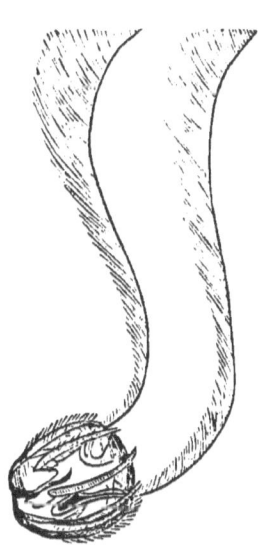

PLEUROBRACHIA (Agassiz).

actually coming in contact with them ; while others are not only gorgeously colored, but of a very definite consistency to the touch. Almost any day at certain hours may be seen stranded on the clean sands at Manhattan Beach hundreds and thousands of these shining little balls of life, varying in size from a pea to a marble. Sometimes so many of these are blown ashore by the ocean-waves that it is impossible for the multitude of people who visit this grand and popular sea-side resort, to avoid walking upon them.

There is one variety known as the *Boroes*, which is of globular shape, and so transparent that whatever creature it may have recently swallowed for food may be plainly seen within this little iridescent sphere. Most of this class are furnished with very fine flexible cilia, through the aid of which they glide like shooting balls of light through the water; their movements are extremely varied and graceful —ascending, descending, rotary, and in elegant curves. This little fairy-like object, however, carries its weapons of offense and defense in the shape of two pendent, hair-like filaments, which, though so fine, are tubes with still finer lateral branches, and with these it seizes and safely secures its prey, apparently also stupefying it as by an electric shock. The boroe is also phosphorescent.

A near relation of the boroe is the *Vetella*, shaped like a Japanese parasol, with a thin filament on the top like two jibs placed with their broadest ends together and united at the centre. This sail-like appendage extends across the central zone of the animal. Its color is dark blue. Its circumference is edged with tentacula, and on its under surface are a number of suckers, with which it may either attach itself to foreign objects, if it wishes to anchor, or secure food for its very transparent stomach.

The general principle upon which all these kinds of animals are constructed is that of a floating bladder, which can be filled or discharged of water at will, with a greater or less number of tentacula and long, stinging filaments, and some appendages as sails with which to trim these bladder-boats, or else cilia to act as oars or means of propulsion.

Among our native hydroids is the *Sertularia argentea*, which is found from the latitude of New Jersey to the Arctic Ocean. It is most profusely branched, and probably grows in larger masses than any other species. Specimens can almost always be found which have been washed ashore, lying high and dry, at Coney Island; and, in this state, I venture to say that not more than one person in a thousand who pick this up supposes it to be anything but a vegetable production—some kind of sea-weed; but it is altogether animal, built up by millions on millions of little hydroid polyps, almost invisible to the naked eye, but developing a world of beauty under the microscope. In its dead and dried con-

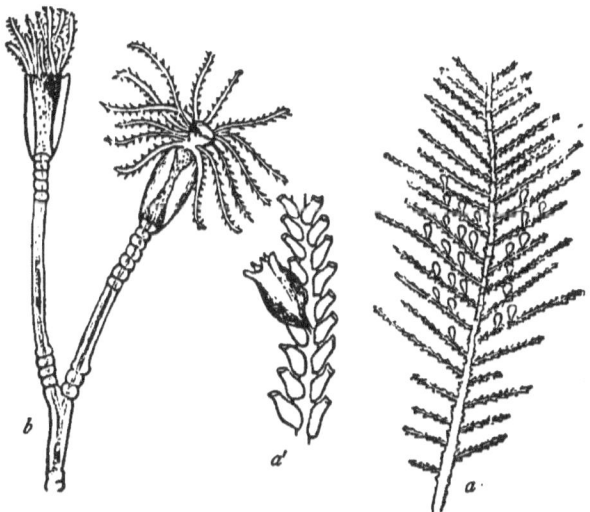

LIVING HYDROZOA.—*Sertularia pinnata:* a, natural size; b, enlarged.

dition it is of so fine and elegant a texture as to take the place of honor among dried ferns, and other delicate plants or algæ, usually without exciting the least suspicion in the

LIVING HYDROZOA.—a, b, Different forms of *Sertularia.*   *Plumularia.*

minds of its preservers that they are carefully *cherishing an animal skeleton.* The *Sertularia pumila* does not grow in such large masses as the former; it may be found attached to the lower sides of stones, or creeping along the sides of fucus, eel-grass, and different kinds of sea-weeds, and is a most beautiful object for the aquarium.

Another of these hydroid communities is the *Coryne mirabilis*, sometimes called *Sarsia*. These are particularly in-

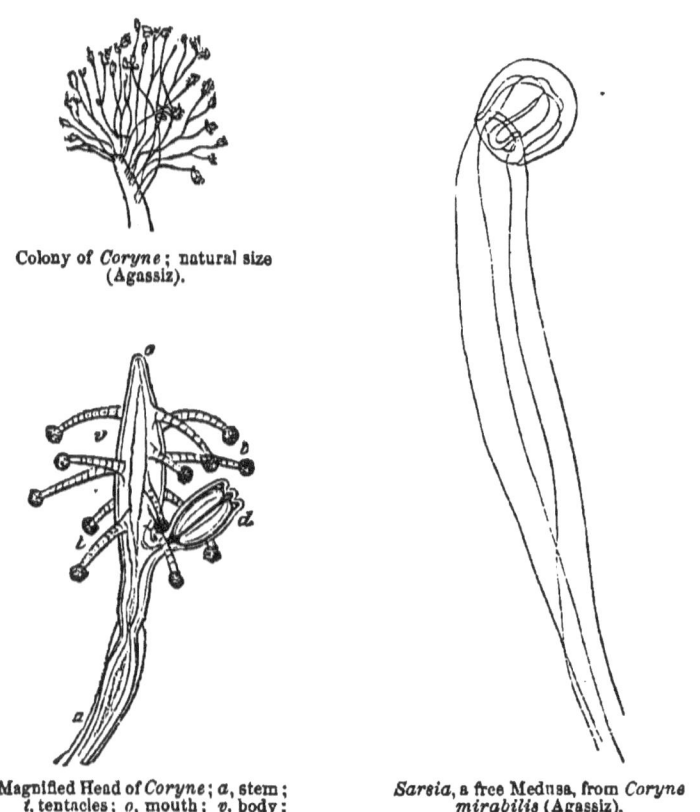

Colony of *Coryne*; natural size (Agassiz).

Magnified Head of *Coryne*; *a*, stem; *t*, tentacles; *o*, mouth; *v*, body; *d*, medusa (Agassiz).

Sarsia, a free Medusa, from *Coryne mirabilis* (Agassiz).

teresting from the fact of their being one of the first species which revealed to naturalists the true relationship of the hydroids to the medusæ. They are very pretty as seen arranged in little clusters on stones near the shore, or oftener on the broad fronds of sea-weeds. Their shape is somewhat club-like, with a small ball terminating each tentacle. They begin to show themselves in April or May, and at this season I have had thousands of these interesting little creatures in a vase of sea-water holding not more than two gallons. I remember well what enthusiasm and perfect delight were expressed by Professor Louis Agassiz when I showed him this *large-little* family of hydroids; he said they were most intensely interesting to him, not only for their intrinsic beauty, but in a scientific point of view.

The *campanularian* hydroids are, like the sertularians, found in miniature communities. The *Tubularians*, on the contrary, are often found extending to a height of four inches; they exist sometimes sin-

gly, and then again in groups containing two or three dozen. The tubularian is ensconced in a tough, horny, semi-transparent tube, about the diameter of a fine knitting-needle; and the little occupant can be discerned through the walls, appearing like a disjointed mass of reddish fluid. The horny sheath does not extend, as does the sertularian's, in a protecting calycle around the head. They have the

SINGLE HEAD OF TUBULARIA, SHOW-ING MEDUSÆ-BUDS (Agassiz).

very singular habit of self-decapitation. A head rises into view from out of the end of the tube, and at about the end of four days *drops off*, when another promptly takes its place and expands for the same brief period, only to undergo the same fate as its predecessor. The opening and expansion of these little heads is like that of the flowering of minute daisies! a wonderful sight, which I hope none of my readers may die without witnessing. In a subsequent chapter of this book I will more fully describe how and where these beautiful objects for the aquarium may be obtained.

Hydractinia polyclina, a velvet-like mass, frequently found on the shells inhabited by hermit-crabs and on the whelk, is yet another species of hydroids, composed of such thickly-massed individuals that with the aid of the naked eye only it would be impossible to count them. These, however, are actually united together at their bases. In clear water, and when fully expanded, they exhibit a reddish or white appearance.

There is another class of minute creatures, a sort of *moss-like* animals, called *Bryozoa*. They are found on marine plants, constructing their reticulated, lace-like cells. They are compound animals, and belong in one aspect to the mollusca family, and are shell-fish anatomically speaking, although they have neither a venous nor an arterial system, the nutrient fluid being contained in the visceral cavity. The *Bryozoa* are found very plentifully on the Gulf weed (*Sargassum*), on little round balls of the size of a pea ; and the " Sargassum " plant is often covered with them. I have frequently fished it up when crossing the Gulf Stream, and the plant was found to be perfectly coated with their coral-laceous cells. On putting a branch of this in a glass containing pure sea-water, the transformation was like magic : immediately this dead-looking object became all life and motion ; their fine tentacles were displayed around each opening ; their form was that of a *perfect wineglass ;* and, as they popped their little plumed heads in and out, moving so lively and even gracefully, I thought it would be difficult, indeed, to see a more beautiful sight. The *Vorticellæ* or tree-like animals are far more unsubstantial than the moss-like, and are fairy-like in the extreme delicacy of their construction. Their frailty compels them to exist as parasites on various kinds of sea-weeds. One might almost compare them to a tiny *drop of smoke or mist :* dew seems too substantial to represent them. They stand out on the least bit of a pedicle, with a little rounded top composed of separate branches, but which looks more like a passing breath than a living animal. Yet how quickly they show that they are alive! At the slightest disturbance every little plume is shut up, and down they settle, so close to the leaf or whatever they are attached to, that they cannot be seen without a lens. But now maintain perfect silence, and keep the water undisturbed, and presently out they come again, each one raising its head like the simultaneous opening of numberless minute web-like umbrellas.

As the subject of *Hydroid medusæ* is almost illimitable, we shall be obliged to give but a brief account of a few more of the most interesting members of the family. Should we attempt to give even a moiety of those which exist, several volumes would be needed; and we are compelled to limit ourselves to a single chapter.

Probably many of our readers who are in the habit of visiting Europe occasionally have had their attention called at some time during the voyage to the presence of creatures called by the sailors "Portuguese men-of-war." They appear only when the sea is smooth, and then they may often be seen, sometimes in large numbers, occasionally "solitary and alone," proudly sailing before the gentle breeze. We say "gentle breeze" advisedly, for at the earliest appearance of "a blow" they reef sails and go below. This is certainly one of the most elegant of the medusa tribe. Its body is of an ovate, somewhat boat-shaped form, with a projection at one end, which might be fancied a little like a bowsprit, only that it points downward instead of upward; it is, in fact, a cylindrical float or bladder, expanded at the centre and tapering toward each end. The most prominent color is a bluish-purple tint, but this changes with every flash of light, so that one can scarcely speak of it as a positive color. For its sail it has a most elegant double frill of white tipped with a bright vermillion or sometimes carmine color. On the under-side its keel consists of a large number of stout, strong tentacles, each of which having a spiral twist, they look on closer examination like ends of blue-silk cording;

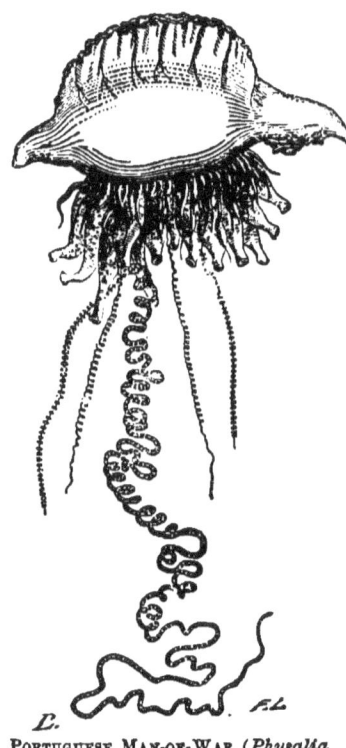

POBTUGUESE MAN-OF-WAR (*Physalia arethusa*).

sometimes these are temporarily grouped together, looking very like tassels. These are not longer when in repose than the depth or vertical measurement of the hull of this fantastic "man-of-war." But several times longer than these are numerous fine, contractile, thread-like organs; while the central streamers extend sometimes to a length a hundred and forty or fifty times greater than the body of the animal. These are sometimes seen trailing at length, and sometimes more or less coiled, glistening beneath the water, spotted with round, pearl-like knobs, making the *Physalia* one of the most attractive objects to be seen on the surface of the ocean. It is a member of the *Siphonores* family, and, like most of the class *Medusæ*, is phosphorescent. Their natural habitat is in southern waters, particularly in the Gulf of Mexico, and those which are accidentally met by European-bound vessels have been drifted out of their course by the northward-moving waters of the Gulf Stream.

Is it not natural to wish to capture such a beautiful object, and to examine at leisure those tender, silk-like streamers? Beware, my friend! Do not too hastily grasp after these ocean-coquettes; all is not lovely that decks itself in gorgeous array. Hear the fate of Father Dutertre, who was one day sailing in the Antilles, and tried to take one of these harmless-looking beauties in his hand. "I had scarcely seized it," he writes, "when all its fibres seemed to clasp my hand, covering it as with bird-lime, and I had hardly felt it in all its freshness—for it is very cold to the touch—when it seemed as if I had plunged my arm up to the shoulder in boiling oil; and this was accompanied by pains so strange that I could scarcely prevent myself from shrieking." Leblond, while bathing in the Caribbean Sea, had one of these animals seize upon his shoulder, and he describes the pain as so intense as almost to produce a fainting-fit. He succeeded in getting rid of the creature, but some of the filaments remained sticking in the skin, which caused great suffering for hours. Many other instances are recorded of the danger attending contact with these elegant sailors, particularly when encountered in their native seas, where they are naturally in the most vigorous condition.

I have had some personal experience with this ambushed warrior, with its concealed mitrailleuse of hair-like guns, which sometimes attain a length of sixty feet, while the body of the animal may not be more than six inches. Though generally represented in illustrated works as trailing, I have seen them coiled up like a spiral spring, occupying but a few inches. While sailing through the Gulf Stream I have captured them with a common hand-net; but in doing this one must be very careful not to let the least particle of these stinging streamers come in contact with the unprotected hands or any part of the body. As soon as these valiant "men-of-war" are thrown upon the deck, they become perfectly helpless; and you will be very fortunate if you find that you have secured them intact; very often, in landing them, these long tentacles, or some of them, are broken off.

Sailors have such a dread of them that one can scarcely procure any assistance from them either in landing or protecting a specimen; indeed, after the creature has been removed and carefully secured in close jars, they will generally wash and scrub the deck, lest some of the "stings" should penetrate their feet.

Captain Mortimer, of the ship Hamilton Fish, has made some most interesting and valuable observations on these animals within the last year (1877). He has even succeeded in preserving one which maintains its sail set, the air-sac remaining inflated, and the iridescent colors preserving all their beauty. Though it had lost some of its tentacles, it was evidently a "Portuguese" of spirit, and not to be dejected by captivity.

The general name of *Medusæ* was given to this class of animals on account of the snake-like filaments which they all possess, and which are highly suggestive of the snaky locks of the Greek Medusa, one of the three Gorgons. And the petrifying power of the latter is practically exercised by their marine namesakes; for if their looks are less terrible, their embrace may prove as fatal. And yet how beautiful they look as they move with a sort of pulsating motion through the water, generally borne by the tides and currents, but appearing to ride them voluntarily! Sometimes hundreds of them may be seen floating along, showing every shade and tint, from the brightest to the most delicate opalescent hues.

How different is their appearance, thus disporting themselves, from the wretched aspect they present when stranded on the sands! But then the fine tissues, which in fact form the framework of the animal, can be examined at leisure; and it will be found that in nearly all there are four elongated oval marks crossing each other nearly at right angles, variously tinted with some shade of red, and which are the seams or lines of juncture of the slight, sac-like skin which holds together this unsubstantial aqueous animal. But the best way, if possible, is to secure the medusa alive if one would see the *modus operandi* by which this slight epidermis and these trailing threads resolve themselves into a beautiful form, which seems to hold the secret of the prism within its dilated cuticle. If then our medusa can be dipped up in a bucket, or some vessel large enough to secure it unmutilated, and the water drained or poured off, in a short time the animal will shrink away to a mere fibrous remnant. If now, before it actually dies, a new supply of water is added, a little at a time, the whole process of distention will be readily seen, and the animal will presently rise with something of that pulsating movement which may be observed in a balloon during the process of filling it with gas. When nearly full it tugs at the cords, anxious to get into the aërial space ; and as the medusa fills all its cellules with the fluid which gives it shape and consistency, it leaves the bottom of the vessel, floats gayly to the top, and once more revels in the air and light of the surface.

It is well that all of the medusæ are not so venomous as the *Cyanea capillata*, not only the most numerous but the most virulent of those which frequent our shores. Bathers ought to know how to recognize and avoid these creatures ; for there is no doubt that to them may be attributed some of the sudden drownings, apparently otherwise inexplicable, which have occurred with expert swimmers when at no great distance from shore. This animal is called in common parlance preëminently " the stinger," and perhaps we cannot do better than to quote the story which a recent writer has published regarding this particular variety. He says :

" One morning toward the end of June, while swimming off Long Branch, I saw at a distance something which looked like a patch of sand, occasionally visible, and occasionally covered as it were by the waves, which were then running high in consequence of a lengthened gale which had not long gone down. Knowing the coast pretty well, and thinking no sand ought to

be in such a locality, I swam toward the strange object, and
had got within some eight or ten yards of it before finding out
that it was composed of animal substance. I naturally thought
that it must be the refuse remains of some animal that had been
thrown overboard, and, not being anxious to come in contact
with it, turned to swim away. While I was yet approaching
it I had noticed a slight tingling in the toes of the left foot,
but, as I invariably suffer from cramp in those regions while
swimming, I took the 'pins-and-needles' sensation for a symp-
tom of the usual cramp, and thought nothing of it. As I swam
on, however, the tingling extended farther and farther, and began to feel very much like the sting of an old nettle. Suddenly the truth flashed across me, and I made for shore as fast as
I could. On turning round for that purpose, I raised my right
arm out of the water, and found that dozens of slender, transparent threads were hanging from it, and evidently still attached to the medusa, yet some forty or fifty feet away. The
filaments were slight and delicate as those of a spider's web ;
but there the similitude ceased, for each was armed with a
myriad poisoned darts, that worked their way into the tissues,
and affected the nervous system like the stings of wasps."

The writer adds that before he was able to gain the shore
the pain had become intense, and increased into torture on
emerging from the water. Both the respiration and the action of the heart were dangerously intensified, and the whole
system fearfully shocked. We suspect, however, that these
latter symptoms might be accounted for on the score of
fright, which such an encounter might readily excuse. The
application in this case of a soothing oil outwardly, and the
free use of brandy inwardly, proved mitigating. But if a
closer contact had occurred, and these long, poisonous filaments had been thrown over the torso of the swimmer, it is
quite probable he might not have returned to tell of the ad-
venture. The concluding sentence may prove serviceable
to bathers in general :

"If the bather or shore-wanderer should happen to see, either tossing on the waves or thrown upon the beach, a loose, roundish mass of tawny membranes and fibres, something like a very large handful of lion's mane and silver paper, let him beware of the object, and, sacrificing curiosity to discretion, give it as wide a berth as possible ; for this is the fearful 'stinger,' the *Cyanea capillata*."

CHAPTER XIV.

To most people "a sponge is a sponge," and "it is nothing more," like the primrose to Wordsworth's countryman; but scientifically a sponge is *an animal*, and it is something so different in its living state from what it is in the dried and withered condition in which it is usually seen that it is no wonder that even naturalists were long in attaining to the knowledge of its essentially animal character.

The sponge as most of us know it, in commerce or on the toilet-table (*Spongia officinalis*), is simply a bushy or dome-shaped mass of dried, elastic fibres, characterized principally by its capability of being softened by moisture, and, as for its appearance, one would simply say that it was a yellowish or reddish brown, highly-porous substance. But let us seek the living animal-plant; and, if not previously aware of the circumstance, we shall surely make a mistake; for it is not here of a dowdy-brown color, but many of them are of a bright-golden yellow, or, more likely still, an intensely bright vermilion, at a distance resembling coral. This is not all the difference: if we have very sharp eyes, we shall see that over the fibrous mass, and interlining all the interstices, is a slight, gelatinous membrane; and, if we add to our eyes the aid of a good lens, we shall see that the whole substance is interpenetrated with fine spiculæ, irregularly arranged into groups, sometimes of mathematical, sometimes of fantastic shape. These sharp, needle-like spiculæ pierce quite through the investing membrane. Then, if your glass be strong enough, you may see slight, narrow, cylindrical or tube-like projections, with a minute round, bladder-like appendage at the end, in which is a circular orifice; sometimes several of these are united, and other variations of form take place. These tubes are not larger than an ordinary bristle. Their use appears to be to discharge the fecal matters which cannot be assimilated from the particles continually floating through the pores of the sponge. Late experiments have proved that these little tubes are sensitive to the touch.

The scientific world is indebted to Mr. Bowerbank for the discovery of *motory cilia* in the sponge, which settled the controversy as to its animal nature; the particular species on which he made these conclusive investigations was *Grantia compressa*. This variety grows at the tide-level, not in deep water. The family name of the sponges is *Porifera*, on account of the porous nature of the substance. It is the smaller holes that are called *pores;* the occasional larger ones, which are usually found on the convex surface, are called *oscules* (little mouths). As the sponges receive their nourishment through the influx of sea-water, all these orifices are connected in the interior with canals leading into each other, and with the efferent orifices. Under an intense light and high magnifying power, the infintesimal polyp has been seen; both cells and polyps "appear as if blown of clear glass; the edges of the cells of the polyp, of the internal viscera, and of every individual tentacle, having the refulgence of polished silver."

In the spring these minute creatures engender germs, round, yellowish-white bodies, which at the proper period burst, giving egress to embryos of an ovoid shape and granular in appearance. These little mites are furnished at one end with vibratile cilia. Swarms of them appear in April and May; after a short period they come to the surface of the water, where millions of them are dispersed by local

currents.   Subsequently they temporarily lose their cilia,
and are then transformed into flat, gelatinous disks; finally,
they become fixed to some rock or other object, and there
develop into the form of the parent sponge.

The sponge family are found widely distributed through
nearly all seas, and appear to attach themselves almost indifferently to any object, but are very frequently found hanging
from the under-side of projecting rocks or in caves, either entirely submerged or visited by daily tides.   A favorite habitat
is the Mediterranean Sea ; they are also plentiful in the Red
Sea and in the Gulf of Mexico and the waters of Japan ;
some are found on the coast of Wales, and a few specimens
grow on our own shores, mostly affixed to shells ; but the
genus containing the greatest number of varieties is found
in the Caribbean Sea.   Those kinds which grow in the waters of our coast, from Massachusetts to New York, are
nearly all parasite upon shells—oysters, mussels, scallops.
The most common of these are the *Flustra foliacea* and the
Alcyonidia oculata.   At the rooms of the Long Island His-
torical Society in Brooklyn may be seen some beautiful specimens of our local sponges.

Sponges have different characters.   The best sponge owes
its value to the resiliency of the interwoven fibre of the skeleton, while others contain so much flinty matter as to be exceedingly brittle, and have scarcely any elasticity.   Of this
kind is the *Spongia fragilis*, which is almost entirely composed of little hollows or cups, nearly as friable as thin glass.
The flint in this class predominates ; in the common kind a
horny substance prevails.

The *boring sponges* usually attack mollusks.   One species,
the *Cliona sulphuriæ* (Verrill), does not limit itself to the
shell, but spreads out on all sides, killing the animal, and at
last dissolving the substance of the shell ; and it frequently
continues its growth, absorbing sand, gravel, and stones, until
it attains over a foot in diameter.   A very beautiful form of

jewelry is indebted to the boring sponge for its attractive appearance. The German and Sicilian agates owe their peculiarity to the presence of fossil sponges; and the chalk-cliffs of England and the chalk-quarries under Paris are largely composed of fossilized sponges. But one class, that known as *Spongilla*, grows in fresh water.

The "Turkey sponges," so called from being gathered near the shores of Turkey and the adjoining seas, are those we mostly see exposed for sale; but three hundred species are already known. All the valuable kinds live in deep water, and are brought up by divers, or occasionally by dredging-machines; they assume an immense variety of shapes, some very curious, like the "Neptune's glove," which bears a striking resemblance to a human hand. Some of them grow vertically, others hang pendent. Some are exceedingly coarse, and are with difficulty freed from calcareous matter;[1] others are fine and soft, almost as smooth as silk. The most common shape is that of an elevated mound.

I have succeeded in keeping sponge alive in confinement for several months; but it does not thrive under such conditions. It appears to require the free flow of the ocean, the constant dashing of the waves, the rising and ebbing of the tides, to content it and insure its growth.

---

[1] It is these calcareous sponges which Prof. E. Haeckel declares are an important link in the ancestral history of *man!*

CHAPTER XV.

THE WHITE WHALE—THE SEAL.

WE do not expect as a general thing that our readers are ambitious of keeping whales in their parlor-aquaria. But as they have been more or less domesticated in our public aquaria and museums, and as I have had a personal acquaintance with one at least of the species, it seems not entirely out of place to contrast this monster mammal of the deep with some of the almost microscopic forms which we have been considering.

From a hydroid to a whale! What an immense leap,

WHITE WHALE (*Beluga catodon*).

and how many thousands of intermediate animals we have been obliged to omit all notice of for want of space! But this is such a complete contrast to the class of marine objects

which we have hitherto described that perhaps some of our
readers will be inclined to charge us with aiming at a "grand
transformation scene" to close the show; for this chapter
must end the descriptive portion of our book, and we shall
only add some concluding remarks on the best way of establishing and preserving an aquarium.

But to return to our whale. This species (which naturalists call the white dolphin) does not grow to the immense
size which some others attain; it rarely grows to a greater
length than twenty feet, while the "true whale," the "Greenland," and the "sperm," often reach to fifty or sixty. But
a few feet more or less will make no difference on paper,
and it will answer just as well for a type of the great *cetacean*
family as any other member; while it has had the advantage
of social intercourse with mankind in our own metropolis,
and also in the learned circles of Boston, where it has been
under daily investigation by the great naturalist Agassiz and
other *savants*, whence it has acquired additional interest.
Whether the white whale himself is duly grateful for all
these *social advantages* which have been forced upon him
may be questionable. Be that as it may—involuntary mar-
tyr though he may have been in the cause of science—the
community at large has been greatly instructed as well as enter-
tained by the enterprise of our fellow-citizens who conceived
the idea of capturing a white whale for exhibition; for the
sight of an object, especially if its habits can be observed and
studied, is worth many lessons from books, however learned
or entertaining these may be.

None of the whale family are fishes. That they were
long considered so, and are still spoken of as such, simply
shows that the external form and the element in which the
animal lives have alone been considered, not its internal
structure and the essential principles of its organization.
Fishes having been probably the first to attract man's notice
among all the dwellers in the deep, every animal which

lived in water came to be spoken of as "fish." The naturalist, however, has other rules for classifying the animal creation, and the latest and best method is that of internal structure and organic function. The whale differs from true fishes in several important particulars. Fishes have no lungs, . and breathe only by aid of their gills. The whale has lungs like land-animals, and a distinct nostril; in addition, to enable it to remain under water for a limited time, it has an air-chest or reservoir, which it fills through the crescent-shaped nostril near the eyes. Again, fishes produce young by means of eggs. Whales are viviparous, that is, bring forth their young alive. The whole fin-system of the fish is lacking in the whale. In fishes the fins consist of a peculiar membrane inlaid with rays of bony spines. The so-called fins of the whale are more like the flippers of the seal tribe, and consist of an extended or protruded fold of the true skin. The axis of the tail-fin is exactly reversed in the whale tribe. In true fishes the tail is set so as to act horizontally, turning only to the right or left; in the whale the movement is vertical, and beats the surface of the waves like a broad-palmed hand, as many a whaler knows to his cost. In diving, ordinary fish sink at a comparatively low angle; the whale plunges down almost vertically. The whale has warm, red blood, and nurses its young with milk, assuming during this operation a vertical position, and holding the young with its fins.

The grade of the cetaceans is indicated by a marked inferiority evinced in the uniformity of the teeth; that is, there are neither incisors, cuspids, bicuspids, nor molars, but all are shaped alike, while some of the tribe have no visible teeth whatever, only rudimentary ones which have never pierced the integumentary covering. The baleen whales, which furnish the whalebone of commerce, have none, and feed on small crustacea, mollusks, and marine vegetation. Whales are capable of affection for each other; they will neither desert their mates nor their young.

All whales have what is called a " blow-hole," that is, an opening on the top of the head which communicates with the mouth; from this they eject air mixed with water, and this peculiarity enables them to be distinguished at a long distance from seals, porpoises, walruses, or other large inhabitants of the deep. Thus, when the lookout on a whaler sees this spurt in the air, he knows of course that it is the object they are seeking; but he never calls out, " There's a whale!" but invariably, " *There she blows!* "

The white whale is distinguished from its larger relatives by being deprived of a dorsal fin. Whalers never call this variety a whale; it is with them " fish," or " white-fish," reserving the term " whale" for the more valuable species.

The difficulty of capturing a whale even twenty feet long is immense, and the expense is also great. It cannot be taken alive by pursuit, and the only successful method is to build a large trap, by sinking long stakes into the mud at low tide, inclosing a space large enough for several whales, and excavating this so as to form a sort of basin. When a school of whales approach the spot at high tide, they do not perceive the shallowness of the water; and the sole chance of capturing them is, that they remain sporting about over the inclosed basin until the tide has retired, when they find themselves struggling in the mud or in water too shallow to float in easily, and they are then approached by their hunters. Now comes " the tug of war." Several men enter the water and endeavor to fasten ropes around one or two of the entrapped animals. The lower the water, the more easily this is done; but generally a very exciting struggle ensues, the whale endeavoring to escape from the barriers which surround him, and the men in their attempts slipping, splashing, sinking in the mud, sometimes knocked over by the plunges of the worried animal, ordering, shouting, and (I fear I must add) swearing. Finally, in a few instances, they have at last succeeded in making leviathan captive.

But even when the ropes are secured around the whale, care must be taken not to injure him by fastening them too tight; they manage, therefore, rather so as to guide him the way he must go than actually to drag him by main force. Withal, speed is necessary; for should the tide return before the victim is secure, he would get away in spite of men and ropes. A necessary part of the capture has been the preparation of a long, strong box, specially built for the purpose and brought on to the muddy flat. This box is carefully padded with whole cart-loads of fucus and other sea-weeds; else the whale would beat himself to death in his struggles from surprise and anger at finding himself in such contracted quarters.

After the men have succeeded in tumbling him in, the question is how to keep the animal alive during its transportation of several hundred miles. One would think he might be towed round, like the Cleopatra needle; but such was not the plan decided upon with the specimens brought to New York. Each one was first hoisted aboard a schooner, which bore the captives to Quebec; from there they were sent by the Grand Trunk Railroad to Portland, Maine, and thence by steamer to this city. The boxes were not more than four feet high or wide, and just long enough for the body. No doubt, when the poor animal found himself thus inclosed, he suspected that a first-class funeral was contemplated, with himself as the principal in the affair; for the box was little more than a coffin. It was not water-tight, and freely admitted air; and a man was kept constantly employed in watering his lordship's " blow-hole," standing by his head and with a dipper keeping up a sufficient supply of water to enable the creature to " blow " if he wished, or at least to breathe freely. If this important organ had been allowed to become dry, the coffin would probably have passed from a fancy to a reality; but not only was this carefully attended to, but the whole epidermis was kept wetted, and his sea-weed bed moist, so that he was cared for as comfortably as the circumstances permitted.

It is necessary only to follow the fortunes of one, as all were treated in much the same way. To Messrs. W. C. Coup, Charles Reiche & Bro., the founders and original proprietors of the Public Aquarium in New York, is the credit due for the successful accomplishment of bringing several of these whales to our metropolis, where so many thousands have been able to see them living, and where their habits could be studied from day to day.

Those brought by these gentlemen, however, were not the earliest captured and kept in confinement. The first one ever captured for such a purpose was secured by Prof. H. D. Butler, who brought it in perfect health to Boston, Massachusetts, where it was kept in an immense glass reservoir, and at a later date was under my immediate charge and supervision. It continued in good condition for more than a year, and became so perfectly acclimated to its new home that it actually showed some signs of intelligence. There was a nautilus-shaped boat made, to which he was occasionally tackled and taught to draw. I fancy he was not very fond of being treated like a draught-horse; for when we wanted him to " hold up " to be harnessed, he just put on speed and went all the *faster* around his glass-walled circle. He would, however, sometimes condescend to take a live herring or a squirming eel from my hand, and then, turning on one side, sail round and look up for more of the same sort; and in other ways he would show that he was really becoming an intelligent Americanized citizen. This creature hardly ever remained still; it appeared to be always swimming around its tank, and ever in the one direction, but varying its speed; and it seemed to find amusement in diving up and down and in splashing the water with its tail, which was admirably formed for the purpose; varying its performances

9

by occasionally spouting a stream of water through its blow-hole into the air.

It is astonishing how tenacious of life is the white whale, and the amount of ill usage and hard treatment it will under-go. Last season one of these animals was shipped in a box *without water* from New York to the Royal Aquarium in London, where it arrived *alive,* but of course not well; and it was not surprising that it resented such treatment by dying four days after its arrival.

SEALS.

One of the most *intelligent* of marine animals is the seal. There are many varieties, but the *Phoca vitulina* or common seal is the kind I am best acquainted with, and with which I have enjoyed a friendly, even affectionate relation. These are not infrequent on the coast of Maine, and occasionally extend their southern wanderings into the latitude of New York. It is sometimes called the "springing seal," from the agility of its movements in the water. It travels in shoals—unlike the Greenland seal, which is only found single or *en famille.* When appearing in our waters, these seals are often mistaken for porpoises; but they have not the roll-ing motion of the latter, and the characteristic springing leap of this seal would be readily recognized by an experi-enced eye.

The seal is a warm-blooded animal—a mammal, in fact, which suckles its young on milk, which sailors sometimes appropriate for their coffee. Like the whale, it is inclosed in a thick case of blubber, so closely adherent to the skin that it is not easily separated. The skin is covered either with hair or fur; the best fur-bearing seals are found in ant-arctic waters. A full-grown seal of this variety will vary from four and a half to six feet in length; it is compara-tively slim in form, and very lithe and elastic in its move-ments, twisting and turning itself in such an astonishing manner that one would think it had no vertebræ. The shape of the head resembles a dog's, with the ears cut off in the latest style of "black-and-tan" coquetry. The most striking feature is the eye, which is dark in color, but large, bright, and speaking in expression—sometimes having a loving or even pathetic look almost human. When on the land, where they spend much of their time basking in the sun on the rocks, they propel themselves forward partly by their strong, muscular tail (or hind-flipper), which acts something on the principle of the blades of a steamboat-propeller, and pull themselves forward and upward by the aid of their fore-flippers, which are broadly hand-like in form. With these forearms or flippers the maternal seal holds its young as a human mother does her child; and there is little doubt but it is this animal which originally gave rise to the legends about mermaids. Seen among partially-concealing waves by the wonder-seeking eyes of unscientific sailors, a female seal

SEAL SWIMMING.

might really be mistaken for a semi-human being. When it returns to the water, the prominent nostrils of the seal flatten out, closing the apertures and effectually excluding the water. Seals live principally on small live fish; and, when aiming for a dinner, with their bright eyes fixed on a victim, they dart upon it with almost lightning-speed.

They are not only capable of being tamed when in captivity, but it may truly be said they can be *educated*. They are extremely sensitive to sounds,[1] and can even be taught to enunciate short syllables. I had one that could say *Pa* and *Ma* intelligibly, and no doubt with longer instruction it might have acquired other words. This one would also play

THE SEAL (*Phoca vitulina*).

a whole tune through on a hand-organ, by holding on to the crank with its right flipper. He could also make as graceful a bow as any lady need wish to receive, and he would "throw a kiss" with his flipper with much more ease and grace than many persons exhibit; but, while throwing kisses to you with one flipper, he would also throw water over you with the other, and expect you to enjoy *that* as well as himself! He would also follow me about like a dog, and was not even dis-

[1] It is mentioned in the diary of Captain Tyson ("Arctic Experiences," page 303), that seals can be attracted by any soft, pleasant sound, whistling, singing, or words spoken in a pleasing tone; and these devices were used by him to procure provisions for his "ice-floe" party.

couraged by a flight of stairs, up and down which he would go to keep me in sight. When I was obliged to leave him at night, or any other time, he would beckon to me with his head and neck "to come back," just as plainly as words could have expressed the feeling; and when he could no longer see me, he would cry like a child.

Sight, hearing, and memory are very keen. They do not readily forget any one who has petted them, nor an unkind word either, and *never* a blow; and, really, if one had the proper accommodations, he could not easily procure a more loving and grateful pet than a bright-eyed, intelligent seal. The accompanying illustration shows a group of them in the act of looking and listening to the well-known footsteps of a friend not yet in sight.

CHAPTER XVI.

EVERY man ought to have a hobby. There is no real life without enthusiasm for *something;* and there is no passion so healthful as that for natural objects. Few visitors who see the metropolis for the first time, unless they are of

FRESH-WATER AQUARIUM COMPLETE.

a very wooden nature, fail to get more or less excited over some novelty which the ever-changing phantasmagoria of city life passes before their wondering eyes. My first youthful passion was suddenly awakened by a gorgeous poster embellished with these words: "Wonders of the Sea!" My imagination was wonderfully aroused, and, with a friend almost as ardent as myself, I started for the Aquarial Hall; we arrived in sight of the place, and viewed with increasing admiration the astounding illuminated canvas which decorated the entrance. We approached—we entered—we saw! I do not know whether it was the curly fish, with its own tail in its mouth, or the eel-like fish in a double love-knot, or the fat prickly-fish, which looked as though he had swallowed a porcupine and all the spines had stuck through the stomach of his entertainer, that first entranced my astonished vision; or whether it was not rather the bluefish which was trying to swallow the goldfish alive, or some now forgotten marvel; but certain it is that such wild exhilaration shot through my veins that in less than fifteen minutes I had already discounted the wonders which were before my eyes, in my exaggerated fancy thought of nothing less than discovering a mermaid's nest, with eggs ready to hatch out the beautiful beings of my over-excited brain.

From fish to fish I traveled, and from those fishes I never stirred for the whole day; and, when finally dragged away from the place by my companion, my first thought was, "*I must have an aquarium!*" At first I wanted one as big as the Central Park, where could be kept every kind of fish I had ever heard of; then, successively, I felt obliged to reduce the size to that of Union Square and of the Everett House; and thought I was very moderate when I had compressed my imagination to the limits of two city lots, and mentally flooded them for the purpose of fish-culture! But, finally, on reflection, it became apparent to my sobered thought that in so large an aquarium as I had been imagining I should not be able to *see* my fish, any more than if I should drop them into the ocean. So away went my dreams, and, in sad sobriety, I at last concluded to content myself with a tank of *the largest size!*

WHAT AN AQUARIUM IS.

An aquarium is any kind of vessel containing water, aquatic plants, and animals, in a living, healthy, and, as nearly as possible, natural condition. The water may be either marine or fluvial, and the choice of this governs that of the plants and animals. A *tank*, which is a word frequently used by aquarians, is any angular, flat-sided vessel with contents as described above. The successful treatment of animals and aquatic plants in the confined space of an aquarium depends entirely upon the close imitation of Nature in that law of life by which an exact balance is everywhere maintained between the supply of oxygen created in water and the quantity consumed by the animals inhabiting it. Without a due regard to this principle, the aquarium, one of the purest pleasures of our modern times, could never have been established.

This principle of compensation was suspected to exist a long time before it was demonstrated by Priestley at the close of the last century. In a French publication in 1778 the subject was also ably elucidated, and the action of the sun's rays in disengaging the oxygen generated in plants was clearly announced. But the application of this principle to the aquarium is of quite recent date ; the year 1850 may be said to have been practically the starting-point, although M. Charles des Moulins, of Bordeaux, France, had made very interesting experiments in the same direction as early as 1830 with fresh-water plants and animals. Mr. Warrington at the first-named date reported to the Chemical Society of London a series of observations on the adjustment of relations between the animal and vegetable kingdoms. In the course of his experiments he placed two small goldfish in a glass receiver, in which there were some earth and sand and a small plant of *Valisneria spiralis.* All went on well for weeks, until some leaves of the plant decayed and made the water turbid. With a flash of thought akin to genius, Mr. Warrington remembered that in natural ponds there were always water-snails, and that these fed on decayed leaves. He speedily obtained several, and introduced them into his tank. The transformation of affairs was rapid ; the snails immediately began to feed greedily upon the decayed leaves, and the water was soon thoroughly purified and restored to its original clearness. The same gentleman in 1852 instituted a similar series of experiments with sea-water, and was equally successful.

About the same time the celebrated English naturalist, Mr. Gosse, commenced tests upon a much larger scale, aiming perhaps less at general scientific discoveries than at the development of certain lines of natural history. To him, his experiments, and his pen, are due the thanks of all lovers of the aquarium ; for he, perhaps, has done more than any other to make it popular. In this country I believe the writer was one of the very first to be inoculated with the aquarial passion—a passion that has grown with time, and has a deeper hold to-day than even in the first period of magnificent visions.

So far as I have been able to ascertain, the pioneer inductor of the private aquarium in this country was Miss Elizabeth Emerson Damon, of Windsor, Vermont; and her first essays were made with the simple apparatus of a *two-quart glass jar*, with a few fish, some tadpoles and snails, and some *Potamogeton* (common pond-weed); but so *perfectly balanced* was this young aquarium with animal and vegetable life, that I fell in love with it at first sight ; and never since, among all the aquarial curiosities which I have possessed, and the thousands I have seen, has there been a collection nearer perfection than that contained in the poor old two-quart jar —an opinion confirmed by nearly a quarter of a century's experience that Nature's laws are unchangeable, and that it is not the quantity but the *quality* which makes perfect. It is to the nice balances of Nature that we owe so much of health, life, and beauty. Sea-water contains about 35 parts in 1,000 of solid matter, which consists of chloride of sodium, chloride of potassium, chloride of magnesium, sulphate of magnesia, sulphate of lime, and also traces of bromides, iodine, fluorine, phosphates, borates, carbonate of lime, and silica, with yet smaller quantities of iron and manganese, arsenic, copper, lead, silver, and gold. When pure it has about eight parts of oxygen and one of hydrogen gas.

On account of the large amount of animal life in the ocean, there would be a deficiency of oxygen, and an accumulation of carbonic acid, if this were not compensated for by the profusion of marine plants. Were it not for these, the carbonic acid exhaled by the marine animals would cause a general condition of *coma* among them. But the oxygen given out by the abundant vegetation keeps the balance just right. It is not alone the carbonic acid which is injurious to animal life everywhere, and particularly within the limitations of a private aquarium; but too much hydrogen is also very destructive, and any decaying body submerged in water gives off various offensive gases, such as sulphuretted hydrogen and other compounds of a virulent poisonous nature. Hence the aquarium must be carefully watched, that neither dead animal nor vegetable matter remain in the tank.

Probably some people look upon the aquarium as a mere pastime, a sort of adult toy. But, properly considered, it is a scientific apparatus, with ever-new questions to propound; each new inhabitant introduced has its peculiar laws of being, and the experiment is often a doubtful one whether this or that new-comer will not disturb the harmony, or attempt the lives, of some members of the family already happily established. The laws of heat and light, too, have to be considered not only with regard to the health of the animals, but to the growth of vegetation; for this may become too rank, or it may be found deficient in balancing the animal life. The natural laws involved in the successful preservation of a

well-stocked aquarium will often tax the knowledge and resources of the ablest naturalist; while yet they may be rendered so simple (if too much variety is not insisted upon) as
to be within the easy management of any intelligent person.

The first necessity of an aquarium is *sufficient aëration*.
This may be produced in three ways: First, by running water; secondly, by means of growing plants; and, thirdly, by
artificial introduction of air. All of these modes are now
used, and each has its peculiar advantages, and each some ob-
jections; circumstances must often govern the choice of
means, but *the thing must be done*, and the water duly oxy-
genated. When once the first principles are mastered, and
reasonable success assured, I know of no attainable pleasure
so pure, enduring, and elevating, as the observation and study
of these natural objects—not dead, but *living*. And what is
of some importance in these days, the pleasure may be made
very inexpensive; and another advantage is, that it is pre-
eminently health-promoting. We all know—business-men, I
mean—how unsatisfactory it usually is " to take a week " for
a vacation, simply from custom or a feeling of necessity,
when one has no special object in view, and knows not where
to go or what to do; and I believe all medical men will agree
with me in the assertion that exercise taken with an object
is very much more beneficial to body and mind than when
undergone in a mere perfunctory manner from a sense of
duty.

THE AQURIANIST'S RAMBLES.

From the hour that you possess an aquarium, assuming
that you have some taste and love for natural history, you
are supplied with a motive for a pleasant ramble whenever
you have an hour to spare. For equipment, you need only a
small hand-net with fine meshes, a glass vial, and a small tin
pail with a perforated lid. With these you are prepared to
hunt along the margin of the nearest rill, or pond, or sea-side
pool. You will always find some new object, animal or vege-

table, for your tank; you will not make half a dozen expeditions without finding forms of life entirely new to you; and probably there are few places in the world more favorable for procuring supplies for the aquarium, at little cost of time or trouble, than the immediate vicinity of New York City.

For the fresh-water aquarium there are the streams and ponds of Harlem, of Staten and Long Islands, and the rivers of Jersey, all of which furnish abundance of living things, beautiful and interesting objects, both plants and animals; while the shores and rocks of our noble bay, the East River, and the Sound, are an inexhaustible reservoir for the salt-water collection. Before you can form any true conception of the wonderful variety of marine animal and vegetable life to be found within a limited circuit, hidden, indeed, from the common gaze, though by no means inaccessible, you must have tried the experiment of looking for it with your own eyes, and not by proxy. A single scoop of the net along the bottom of the streamlet, or just below the surface along the margin of the pond, will often bring living forms to light which were " never dreamed of in your philosophy."

HOW TO START AN AQUARIUM.

I would advise no one to commence the keeping of an aquarium upon a large scale. A small beginning is safest; for, although you may think you have studied the subject thoroughly, and the perfection of an aquarium is admitted to be governed by fixed laws, yet success can be practically attained only by experience. A fixed shape is not necessary. Even a glass globe, the most objectionable of all, can be used; but it has the serious disadvantage of optical distortion; and, worse yet, if brought into the sunlight, it is acted upon like a burning-glass. I have seen fishes almost parboiled in a globe, which would have been comfortable in the same position in a square tank, or in a flat pan with plenty of surface exposed to the air. Almost any glass vessel can be used, but the vessels made expressly for the purpose are much the best. And I may here mention, for the benefit of possible readers outside of New York, that everything necessary for the stocking of an aquarium, as well as the reservoir itself, can be ordered from the establishment of B. Greenwood, No. 11½ College Place—tanks, glasses, plants, and animals for salt or fresh water, in almost endless variety. Indeed, the furnishing of aquaria has now become an important commercial interest of this city.

The oblong tank is undoubtedly about the best shape, but this rule need not be rigidly adhered to. It must be remembered that, whatever shape you decide upon, there must be a proportionally large amount of *surface* as compared with the depth : this obviates much care as to the oxygenation of the body of the water, which is of course a vital point. You may test this if you have an invalid fish : if he is not too far gone for convalescence, put him into a common earthen dish or pan with fresh water, and he will speedily recover. I have tried this often very successfully with invalid fish.

A home-made tank will do, then ? Don't think of it for an instant as a permanent achievement. My dear reader, you cannot anticipate the difficulties ; you had better attempt to build a house, or a railroad with all its rolling stock, than to construct an aquarium which *will not leak* on the parlor carpet, or kill the fishes it contains. " It looks simple enough," you will say ; but it is *not*. The amount of latent difficulties inherent in the construction of a " home-made aquarium " is perfectly astonishing. You think you have got it all right, and now that it really is finished you invite a few favored friends to be the witnesses of your success ; and as you stand by your crystal pond, one of your friends confidentially whispers to you, " It's leaking, and will spoil your carpet." You knew this well enough yourself ; that is why you stood so *very close* to that particular corner, and tried to conceal the wet spot with your shadow, hoping your guests would not see it.   Oh, dear ! it is too bad, but water is *so thin,* it will work through anywhere.   It drips down faster and faster, and you are obliged to ask your friends to assist you in emptying your grand new aquarium.

I think I hear some reader ask : " Did that ever happen to you ?   And if it did, could you not repair it ? "   Didn't I try to, over and over again ? and sometimes whole days would pass without a drop of water escaping from my tank.   But at night one of the glasses—*only* one, perhaps—would crack its whole width, and the water would flow out so noiselessly upon the floor that the mischief would not be discovered till morning.   Then imagine the consternation and disgust !   But I will leave this (to me) painful part of the subject ; only, by way of warning, giving the items of expense which my first tank cost me :

Paid for services of a mechanic making pattern for iron corners.......	$2 50
" " " of same man fitting aforesaid.....................	1 00
" " three days' work, wood bottom, etc......................	3 75
" " twenty pounds iron castings, corners, etc..................	2 00
" " four pieces of glass...................................	3 50
" " cement which contained white-lead and killed the fishes......	40
" " another cement which was *not* water-proof................	75
" " another not so good as the first......................	50
" " materials for experiments in cement-making...............	2 50
" " a *good aquarium-cement,* and expressing the same from New York......................................	1 00
" " new glasses in a tank, three having split from the *warping* of the wooden frame...............................	2 62
" " express charges on a " wooden fish " sent me by waggish friend (made from a shingle !)............................	50
" " small boy helping me fill tank with water thirty or forty times.	3 00
" " same for emptying leaky tank.......................	2 00
My own winter services, *five months' hard work—*say...............	1 00
Thus far actually paid out.................................	$27 02

And *such* a sight to show for it !   I cannot describe it,

and only wish I could present my readers with a sketch of it
as it looked the last morning before I concluded to go to New
York and purchase a tank which would *hold water*. This
I at last did. Add the price of this (only ten dollars), a complete iron and glass octagon tank, to the above, and we have
a total of thirty-seven dollars and two cents as the cost of
my first aquarium, twenty-seven of which I might better
have saved, with all the fuss and trouble that it cost me in
attempting to compete with the professional dealers !

Having thus given the reader a hint how to procure (and
also how *not* to procure) an aquarium, I will now give some
information as to the best mode of stocking a tank. In the
first place, it is always prudent to *test* the tank by filling it
with water, if only to see that there is no leak or any other
defect in its construction. If it appears to be all right, still
do not be impatient to get your stock in ; let the water remain two or three days, so as to thoroughly draw out any
impurity or taint in the cement. If at the end of that time
you should see an appearance of the prismatic colors on the
surface, there is something wrong; it must be emptied and
refilled, until the water remains clear and pellucid, when it
will be safe to begin operations—never, however, using the
water which has stood in the tank, but emptying it either with
a siphon or by dipping and filling it with fresh. It is not
necessary to put earth at the bottom, as was formerly done ;
it makes the water muddy and does no good. The plants
you should introduce will not need it, as the true aquatic
vegetation draws its nourishment from the water alone, and
only uses the ground for anchorage. A few inches of pebbles on the bottom, and some pieces of rock will answer
every purpose, and do not discolor the water ; they can also
be arranged picturesquely and add to the beauty of the aquarium.

The kind of water for a fresh-water tank is not material ;
for, if the plants are of the right sort and in good condition,

they will soon regulate the water. It is a good plan to let the plants adapt themselves to the tank by leaving them undisturbed for a few days before introducing any animals, especially fish, as they sometimes rush about rather wildly at first when introduced to a new domicile. Besides, it takes some little time for the plants to thoroughly aërate the water. You may know when this is effected by seeing little bubbles of air ascending through the water. When these are visible you may safely begin to stock your tank, taking care *not to overcrowd* it with animals. This is the greatest temptation that besets amateurs.

It may be well to suggest also that you should not put in too much rock-work. Of course a large piece of rock displaces a certain amount of water, and the water is more valuable for the preservation of animal life than rock-work. A piece here and there may be admitted, arranged to form a bridge, perhaps, or a cave, but only to suit *your own taste and fancy.* If the plants grow well, they will afford all the shade that is necessary for such animals as seek it, while it leaves them more space in which to disport themselves. Plants are also pleasanter to the eye; and remember, *rock has no aërating qualities.*

Contrary to the generally-received notion, the algæ and not the higher orders of vegetation (with a very few exceptions) are the most effective aërators even of the fresh-water aquaria, and they are more permanent. I have little doubt that many failures with fresh-water tanks are owing to the desire to cultivate the more showy but unsuitable forms of vegetation. Most of what are called the higher forms of plants, particularly flowering plants, have their seasons of decay, when of course they become a cause of impurity in the water; and deposits of decomposed matter mar the beauty of the aquarium, and threaten the health and life of the inhabitants.

A long and varied experience enables me to speak with

confidence on this matter. I have occasionally devoted a tank to special experiments with vegetation, with very satisfactory results. My object has been principally to ascertain what fresh-water algæ and other plants of a low order would spontaneously appear under the influence of a strong light. The tank upon which I experimented held between fifteen and twenty gallons of water. After washing the pebbles and the glass sides and ends with boiling water, so as to destroy any germs that might remain from the plants which once occupied it, I refilled it with pure water (putting no plants in), and set it in a strong light with a southern exposure. But a few days had elapsed before minute spores—unseen in the water without the aid of a strong glass—had fixed themselves upon the sides of the tank and upon the rocks and pebbles on the bottom. A week or so more, and we had a perfect carpet of green, coating the inside of the tank with its soft, silky appearance. It is almost a perfect aquarium in itself, and handsomer far than one would think from the mere description.

I allow, as is the better way with all fresh-water aquaria, the confervæ to form freely on the two ends and the side nearest the light, keeping the side facing into the room clear, by the application every week of a piece of woolen cloth fastened to the end of a flat rod or stick. And, by-the-way, these confervæ will form in any healthy aquarium, but the glass can be kept transparently clear by the above process. They are somewhat difficult to remove if allowed to grow too long, but in that case do not try to remove them with sand or anything of a harsh nature, for you will thus ruin your glass by scratches, and *they* cannot be got rid of by any means ; the woolen cloth is better than anything else. No attempt should be made to remove them from the rocks or pebbles ; they often produce a very beautiful effect, and do no harm. If this green matter forms in excess by growing too fast, it can readily be checked in its growth by diminishing the amount of light admitted to the tank. The easiest way of accomplishing this is to use a screen of *yellow* paper, as this color is a non-conductor of light.

AQUATIC PLANTS FOR THE AQUARIUM.

The common pond-weed (*Potamogeton*) has been much used for the purpose in this vicinity; but it is not really so desirable a plant as many others. When in a healthy condition, it is beautifully tinted, its hues varying from a lively, pale green to a bright, reddish brown. Its form, too, is graceful. But it will not thrive for any considerable length of time; and, what is more important still to remember, its oxygenating powers are much less than those of other varieties of this family of plants which I shall describe.

Potamogeton crispum is also a pond-weed, and in the water is a very pleasing object, having light-green leaves, corrugated or curled; its general effect is pretty in the tank.

Potamogeton . densum is still another plant of this class, and is far more desirable than the former. It has a bold, fern-shaped leaf, is very ornamental, and pretty hardy; and I would advise its use when it can be obtained. I have found it more plentiful at Fresh Pond, near Boston, than anywhere else ; but have also found it in small quantities in the Passaic River, near Paterson, New Jersey.

Potamogeton heterophyllum and *P. natans* are also found in many of our fresh-water ponds and rivers. They are, however, coarse, large-leaved plants, and, though living always submerged, are not particularly desirable, though they will live and thrive in the aquarium.

Antipyretica gigantea and *A. fontinalis*, commonly called "fontinalis," are moss-like plants, of dark and light green shades, growing on old decayed logs, sticks, and stones, on the bottom of clear-running brooks, and sometimes in cold springs and wells, attached by a hair-like fibrous base, having nothing like a root. They bear a fine, feathery foliage, sometimes growing to a length of two feet. I have found *Antipyretica* high up on the mountains of New Hampshire, on one occasion at Grantham, at a height of two thousand feet above the sea-level, and growing perfectly luxuriantly by the side of the mountain-rills. At these elevations it is sometimes left high and dry during a season of drought; but such is its tenacity of life that, instead of giving up and dying, it seems scarcely to complain, but lives on, defying the scorching rays of the sun. To be sure, it turns rather gray under this *régime;* but as soon as the cooling waters of the brooks resume their course, this much-enduring plant gets back its bright color, and continues to thrive as if evil vicissitudes had no power over it. Nothing can be more desirable for the decoration of a tank than masses of this bright, cheerful-looking plant. One variety of it is quite fine and the other large and coarse, thus making a contrast of both texture and shade, while its oxygenating qualities make it almost *invaluable.* It can be relied upon, too, through the whole year, which is a princely quality for an aquarium plant.

Among the scores of fresh-water plants which I have had in my aquarium during all these years, there is no other which has proved nearly so satisfactory as this. The milfoil (*Myriophyllum spicatum*) perhaps comes next as a successful grower and aërator. It is certainly very showy when growing well in the aquarium, and in shape much resembles a full-rounded plume; but it dies away during a part of the year, so that it cannot be always depended upon. It usually grows to about a foot in height, and can generally be found in the ponds on Staten Island and New Jersey. *Myriophyllum verticillatum,* or whorled water-milfoil, much resembles the last-named species, though rather finer, more compact, and rounded; it is not nearly so plentiful.

Hornwort (*Ceratophyllum demersum*) is much like the

milfoils, though perhaps of a finer foliage, and grows much longer. It is a very desirable plant to have, but is not so easily found. I have never seen it growing naturally anywhere except in the Passaic River near Paterson, New Jersey; but I have understood from Mr. Greenwood, the aquarial merchant, that he obtains it in large quantities from Philadelphia.

Many persons prefer *Valisneria spiralis*, or tape-grass, to all other plants for aquarial purposes. I like it much, and its growth and habits are very interesting to study. It rarely fails to take root and thrive in the tank. In the aquarium it is generally propagated by offshoots, which push forward among the pebbles at the bottom of the tank. Perhaps the

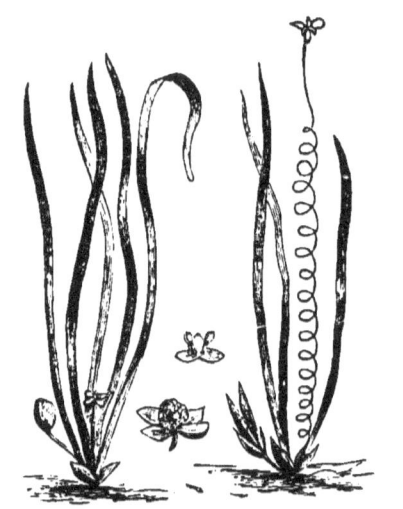

VALISNERIA SPIRALIS—STAMINATE AND PISTILLATE.

most interesting feature in this plant is its mode of flowering. It bears both male and female flowers, which develop from separate roots. The solitary female flower is borne on a long, spiral-shaped stem, which reaches to the surface of the water, and there floats until the male flowers, which grow on a short, thicker stem, only reaching upward a few inches from the bottom, fall off, rise to the surface, and with the diffusion of their pollen fertilize the pistils of the female flower. The latter then coils up its long, spiral-like stem and draws itself beneath the water to await maturity, and in due season sows its ripened seeds. This plant is also interesting to microscopists, as showing very clearly the circulation of protoplasm within its cells.

Ranunculus aquaticus, or water-buttercup, is not wholly aquatic, but has been used quite extensively for aquarial purposes. It looks well, and even ornamental, but is not enduring.

The narrow-leaved starwort (*Callitriche autumnalis*) is readily found in small, sluggish brooks and ponds, and is very pretty in the tank, its pale-green leaves forming a crest of stars on the surface of the water. But it is short-lived, and is liable to be stripped of its delicate leaves by the fishes, and thus disfigured, if not destroyed.

Anacharis Canadensis is one of the best of plants for the aquarium, and grows profusely in many of the small running streams of New Jersey. It is probably one of the most luxuriant growing aquatic plants in the world. It is even mischievous in the rapidity of its growth, by choking and obstructing rivers and canals. Since its introduction into Europe it has increased so rapidly in several localities as to obstruct navigation, and cause considerable expense for dredging and other mechanical appliances to exterminate it, or in attempts to reduce its exuberance.

The white pond-lily (*Nymphæa odorata*) is almost useless as an aërator, but I like it very much for the larger tanks. Its ample, rich green leaves floating on the surface of the water, afford an excellent shelter and hiding-place for many little animals, which would otherwise fall a prey to the fishes. Its beautiful, pure white efflorescence is known to almost every one, and remembered gratefully for its delightful perfume. Though prizing it highly, I have not succeeded in bringing this plant to perfection in my own aquarium. I have, however, seen a wax flower substituted for a natural one, and floating among the lily-pads, so apparently at home

WHITE POND-LILY.

that some excellent botanical connoisseurs among my dearest friends did not discover the deception until they endeavored fruitlessly to inhale its fragrance !

The *Calla*, or common white lily, is more ornamental than useful; but it may at discretion be arranged in the cen- tre of a *large* aquarium (none of these non-aërating plants should be tolerated in small tanks), where its dark, branch- ing leaves look very handsome raised above the water. I have scarcely ever brought it to flower in my own tank, but my sister, Miss E. E. Damon, just informs me that she has succeeded in making one bloom in her aquarium repeatedly —each season for several years. But if one *has* a calla lily in the tank, it is useless to attempt to keep newts or tritons, as they will invariably climb up the stems, and end their career by falling over the side of the tank, where fatal acci-

CALLA ÆTHIOPICA.

dents are pretty sure to overtake them before they can be restored to the water—especially if any representative of the feline race should happen to spy them.

Lymnanthemum lacuosum, commonly called ox-heart, so far as I can ascertain, is to be found exclusively in a small pond at Paterson, New Jersey. The leaves are a pretty shade of bronze or dark green, about the size of a silver dollar, though heart-shaped; it floats in small clusters on the surface of the water, and from the little bunch of leaves a small white flower holds up a tiny, smiling countenance. Its mode of growth is something like that of the strawberry-plant, putting out a runner from the central nucleus, and at every few inches of departure developing a new cluster of leaves and roots. It looks lovely in the aquarium, and I always consider it a special treat to possess a plant of it.

Calamus, or common sweet-flag, is not one of the useful

plants in the tank; but it grows well, and, if there is room for it, *looks* very well indeed, shooting up its straight, spear-like branches. Put in a corner it will do no harm.

FLOATING VEGETATION.

Naias flexilis, or *Nitilla*, is one of the prettiest of the floating species of water-plants, and is also otherwise desirable for the tank. I should always endeavor to include it in my collection. It grows in masses, looking something like a snarl of hair of a bright emerald green. It is, however, branched at the joints, and is a favorite subject for the microscope on account of the visible circulation of its protoplasm.

Riccia natans is in general appearance not unlike green sawdust, but on closer examination the shape of its leaves is found to be nearly triangular, being formed of three lobes branching from the central axis. I like it for the tank.

Lemna purpusilla and *Lemna trisulca*, two more floaters, known as duckweeds, are also good plants for the aquarium ; for, besides their pretty appearance, their myriads of small floating leaves harbor thousands of minute living beings—small snails, and so forth—on which the larger animals may breakfast.

Water-net, a plant of which I do not know the botanical name, is a beautiful growth, and well repays a long tramp to procure it. It is only to be found in still ponds, and may be distinguished by its light-green color and fine hair-like fibres, perfectly woven into finer meshes than any fishing-net which can be bought.

Of the *confervæ* which grow spontaneously, or can be introduced into the aquarium, there are several desirable species. They are good aërators, and often very elegant in appearance, sometimes stretching in graceful festoons from plant to plant, or forming an emerald arch across the tank. Among the list of those I have had are *Cladophora*, *Hyalotheca*, *Draper-naldia glomerata*, *Drapernaldia nana*, *Tetraspora ulvacea*,

Hyrodyctyon utriculatum, Batrachospermœ, Tyndaridia pectinata, Zygnema inequale, Ocillatoria pulchella, Lyngbya virescens.

One most emphatic word of advice to the beginner is this : When your rocks are once arranged, *let them alone.* This advice is not new or original, but it cannot be too often repeated. You might as well go into your garden and pull up your plants to see if they are growing as to attempt to rearrange your aquarium after it is once stocked. Aquatic plants are just as sensitive to such disturbance as the choicest plants of your conservatory. Do not be discouraged if they look drooping for a few days; still we repeat, *Let them alone.* It is not to be expected that they will look fresh and bright while they are busy at work with their roots, trying to find the fittest place to anchor among the pebbles. And even if a main stem dies from the shock of removal, or fails to become acclimated, have patience ; do not pull out the whole plant ; *wait a little longer :* very likely new stems will spring up, while over the old ones confervæ will grow, speedily covering them with a delicate moss, which will assist in oxygenating the water. I have had tanks, the contents of which, both animal and vegetable, have *not been disturbed for years,* the plants and animals being all maintained in a perfect condition of health.

ARTIFICIAL AËRATION.

Different contrivances have been employed to effect the artificial aëration of aquaria ; and also to keep up a supply of running water, which imitates most closely the natural method. This can be readily accomplished in a dwelling supplied with pipes from an elevated reservoir. All that is needed is to lead a supply and waste pipe through the tank. Where this mode is adopted, there can be little danger of disease or death through lack of oxygen, unless the tank is unreasonably crowded. But such an arrangement will

10

scarcely satisfy the aquarial student; for, however effective, beautiful, *and expensive* this system of aëration may be, *this is not strictly an aquarium.* It is not based upon an equilibrium or balancing of natural processes, which alone constitutes a scientific arrangement, and adds the charm of a successful manipulation of natural forces under artificial conditions—a triumph of which any amateur aquarian may be proud. I also doubt whether the more delicate forms of vegetation would be developed in a tank supplied with water through pipes.

Probably the best form of aëration is that now extensively used at the New York Aquarium, namely, the introduction of air into the tanks by aid of a steam air-pump, led through India-rubber tubes concealed among the rocks and pebbles. The effect to the spectator is very beautiful, as the air thus introduced at the bottom of each tank rises in the form of silvery globules through the water. For large public tanks this is an excellent plan; indeed, the larger fish could scarcely be preserved without either that or running water; but it is obviously not adapted to private dwellings.

For " your own " aquarium other means must be sought; and the *best of all means,* the true scientific and natural mode, which the Creator himself has adopted, *is to oxygenate the water by plants.* No other method will keep it so clear. As evidence of this, I can to-day take a tumbler of water out of a tank that has had only the natural aëration of plants, which has not been meddled with for years, and in which I have kept all that time a large and varied assortment of animals; and it will equal the Croton in purity and clearness, and *far surpass it* in softness, and in *living, sparkling brightness.*

REPLENISHING THE AQUARIUM.

The vegetation in the tank being plentiful and healthy, as will be shown by its sending up innumerable small globules of air, the next thing is the selection of animals—the necessary first, the ornamental next. The reader is particularly advised to avoid the common error of selecting *too large fish*, or too many. It is of the first importance to supply your tank from the start with *fresh-water mollusca and crustacea*. Snails are almost a necessity. To obtain them, let us take a walk to the nearest rivulet. Brush your hand-net along the margin, grazing it against the overhanging grass and weeds. How readily you have obtained a number, but perhaps all of one kind! You "thought all water-snails were alike." In that you find you were greatly mistaken; and this is your first practical lesson in natural history. The snails you have probably got are called *Physæ*, and are of all others best adapted to your purpose. They will consume the conferva which forms in the tank, mowing it from the sides in swaths, as the mower's scythe does the grass. There is but one drawback to them : their delicately-mottled shells are so fragile that they easily fall a prey to any of the larger fish which may be in the aquarium.

Another kind is the *Planorbis* or trumpet-snail, which is also ornamental, being in form similar to the ammonites, so remarkable among the shells of a former epoch. The general belief is that it will destroy the plants ; but my own observation does not confirm this bad character. It is not often found in running streams, but is abundant in ponds, ditches, and low marshes.

Next may be mentioned the *Lymneas*, with a sharp, spiral, intensely black shell, which is the most ornamental of all. I cannot say much for their industry, however ; they enjoy rest too well. The *Palludinæ*, another variety, are useful, but there should not be too many of them.

Some of the bivalves of the fresh-water mussel tribe may be introduced for variety, and have been growing in favor as cleansers. The "swan-mussel," for instance (*Anodon cygneus*), is an interesting object on account of its large size ; but a handsomer kind is the *Unio radiatus*, which abounds in the Passaic River and the ponds and streams of New Jersey. Some unios are rainbow-tinted, and others are of a beautiful rich green. But perhaps you will say, " They are too inert for an aquarium." There again you are wrong, as close observation of your tank will prove to you. They will sometimes travel the whole length of the tank in a few hours. While I write, one of them has just turned itself over with a jerk, and (by-the-way, it is " moving-day ! ") is evidently preparing to shift its quarters, proving itself a true New-Yorker.

Having several times referred to the Passaic River, I will here mention that I have there found adhering to the *Valisneria*, which is abundant in that river, a beautiful, small univalve, the scientific name of which I do not know.

The crayfish or crawfish (*Astacus fluviatilis*) is frequently but erroneously called the fresh-water lobster. It is a very

CRAWFISH (*Astacus fluviatilis*).

interesting object in the aquarium, but cannot well be kept in a deep tank without elevated rock-work, by which it can ascend nearly or quite to the surface. It affects shallow brooks, and can be found in such localities on Staten Island.

With less trouble it may also be procured from the well-known fish-merchant, Mr. Eugene Blackford, of Fulton Market, on whose marble tables specimens can be seen by thousands. Small ones should always be preferred for the tank: if too large, they will disturb the plants and loosen them from their anchorage. The crawfish is very fond of working among the pebbles at the bottom of the tank, and will build itself a cave most ingeniously—lifting stones much larger than itself in its claws, and with them building up a wall with great regularity in front of its hiding-place. From what I have seen in my own aquarium, I believe the craw-fish frequently leaves the water at night. Perhaps it goes on foraging expeditions; for through the night, at almost any hour, they may be found on the dry rock in the centre of the tank, where they rarely appear in the daytime. It is altogether best to give them a separate tank if possible, as they are apt not only to commit depredations on the fish and plants, but to disturb more or less the general arrangements of the aquarium.

FRESH-WATER REPTILES AND INSECTS.

" Oh ! there is a lizard ! " How many times have I heard this exclamation, when only an innocent little triton or water-newt made its appearance ! True, they are somewhat lizard-like in form ; but the lizard is wholly a land-reptile, while the newt delights in the running brook. They are entirely innocuous, and very curious and pleasing inmates of the aquarium. Nothing can be more eccentric than the perform-ances of one of these little fellows. He is the gymnast and acrobat of the tank ; his antics exceed belief. Now we see him poising himself by one foot on a leaf, now " treading water " as bathers sometimes do ; then again suspended, mon-key-like, by his tail from the stem of a plant ; then, perhaps, sitting upright like a kangaroo at the bottom of the tank ; next, darting hither and thither in a frolicsome mood ; so that the eye never wearies of watching him. If he could only speak, we should put him on a salary as " clown " to the concern. But he is useful as well as entertaining, feeding upon the minute parasitical insects which injure aquatic plants. His progressive development is also curious; he goes through eight or nine transformations of form before his latest and most perfect is acquired. No aquarium can be complete without two or three of these lively and amusing creatures.

The frog, being amphibious, cannot well be kept in the tank, unless you have a projecting rock or floating island; and then it is apt to make a bold leap for liberty, and if an open window is near Mr. Frog may have utterly disappeared, without so much as saying " Good-by."

The cleansers having been duly located to keep the apartment in order, like the *neokoros* of the famed temples of Greece, the fish may be introduced; and it will be a pleasure to me to make you acquainted with the

FINNY DENIZENS OF THE AQUARIUM.

Of these, perhaps the stickleback (*Gasterosteus*) is the most interesting; and, although I presume most of my readers

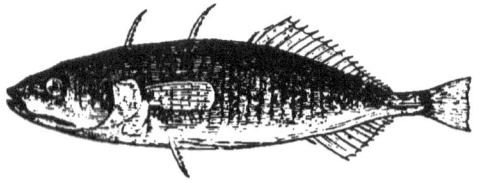

TWO-SPINED STICKLEBACK (*Gasterosteus biaculeatus*).

have already read accounts of his doings, I cannot altogether omit a description of him, without leaving these pages incomplete. There are several species of sticklebacks, varying in size and form, and in the number of their spines; but they are all pretty and interesting creatures, and well suited for the aquarium. The smaller kinds abound in the creeks and ditches of the Jersey meadows, and the larger kinds (which will breed freely in the tank) are found in the shoal waters of the East River and in the creeks on Long Island. They are all, the latter especially, pugnacious little fellows, and quickly become the tyrants of the miniature ocean of their prison. They prefer brackish water, but will live in fresh; the small-tailed or nine-spined variety is least adapted for the latter.

In the breeding-season the male stickleback is more than usually disputatious, and will successfully assail fishes five times his own size and weight. At this period, too, he assumes colors more gorgeous than my pen can paint, while his little eyes glitter like emeralds. When building his nest, which he does with all the care and precision of the most skillful bird, he is peculiarly excitable. He is an aquatic representative of Mormonism, being a decided polygamist. When his nest is completed, both an entrance and an exit are provided for. Into the former he *drives* the female; and, at that moment, so intense is his excitement that his brilliant hues forsake him, and he absolutely turns pale, or silvery white. The spawn and milt are both deposited in this tunnel-like nest, as the male and female fish pass through the aperture. Then the male takes up his position at the entrance, "standing guard" with most praiseworthy perseverance; and by a rapid motion of his pectoral fins, backward and forward, a perfect current is created, which flows through and over the eggs. Over these he keeps vigilant watch for ten or twelve days; there are hundreds of them in one nest, and were they all pearls they would not be so precious to our little stickleback as are these small glutinous masses from which the young fish will now appear. But his fatherly care is not yet ended; for eight or ten days longer he continues to watch with the most intense solicitude, keeping them from straying, which should any of them

STICKLEBACKS BUILDING THEIR NESTS.

do, he may be seen *taking them up as carefully in his mouth* as a cat would her kittens, and restoring them to their home in the nest. The breeding-season of the stickleback begins in January or February. To raise them in confinement, it is much the best to give them a tank to themselves. The male is very ravenous, his food consisting mainly of small insects, chrysalids, and young fish emerging from the spawn. He will also take worms or a bit of meat *scraped very fine.*

Gold and silver fish live well in the aquarium, but the smallest size should be selected for the purpose; large fish will not thrive unless the tank is artificially aërated.

The white perch is a very beautiful inmate of our parlor ponds. The same rule about selecting the smaller specimens should be observed here also; and this general maxim in regard to fish cannot be too strongly impressed upon the amateur aquarianist. The white perch is plentiful in the Passaic, and can be easily tamed so as to feed from the hand, as indeed can most fish. The perch likes worms. There are other varieties of the perch family, all beautiful, but somewhat dangerous company for other fishes.

AMERICAN BREAM OR SUNFISH (*Pomotis vulgaris*).

You would better avoid the common sunfish, unless very small. He is beautiful, but an assassin of most murderous habits. Have nothing to do with him, notwithstanding his beauty. There is another variety of sunfish, however,

called the rock-sun, which is an excellent inmate of the tank.
He is richly marked all over with gold, silver, and bronze,
and is of a most harmless disposition; one could scarcely
believe that he belongs to the same family as his namesake
described above.

Of the dace tribe, there are the red-fin, the white-fin,

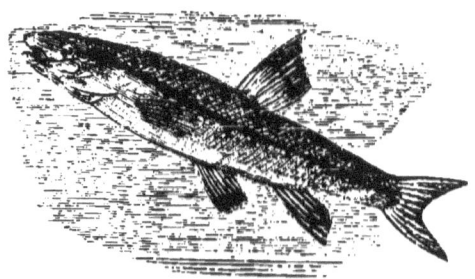

BLACK-NOSED DACE (*Argyreus atronasus*).

and the striped—all desirable fish for the tank. The pike
if very small—not above two inches in length—will do; but
beyond that size he is a dangerous fellow.

COMMON PIKE (*Esox reticulatus*).

Eels are ornamental when small, their wavy lines and
undulating motions offering a pleasing variety; but they are
very destructive of snails, and are better dispensed with.

A very pleasant and lively little fellow is the common
minnow; and the "barred killia-fish" is exquisitely beautiful as well as active.

Among other suitable fishes is the rock-fish. It has a
beautiful and graceful form and curious geometrical markings; and, on account of certain peculiarities, the unusual

shape of its caudal fin, and other points, it was a particular object of scientific interest to the late Prof. Agassiz.

The black bass (not the lake-bass) is a neat, bright, harmless little fish, always improving on close acquaintance.

The common catfish is not amiss, if he is not permitted to grow too large. His bearded chin and curious form are very attractive when he chooses to show himself. In the

CATFISH OR HORNED POUT (*Pimelodus atrarius*).

daytime he is apt to hide himself among the rocks; but he never fails to come out on his evening prowl for food.

The "tessellated darter" may be admitted. He is a sly fellow, and will lie on the bottom of the tank as if fast asleep; but, if any suitable prey is passing, out he darts with a sudden spring as lively as if sleep was a thing unknown to him.

The small-sized trout is very beautiful, but delicate for the aquarium; and, if he does live, he soon proves dangerous.

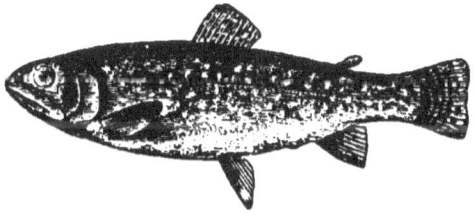

SPECKLED TROUT (*Salmo fontinalis*).

The tadpole, though not belonging to the "finny tribe," may as well be mentioned here. He must by no means be omitted from our collection, for he is one of the "funny" little creatures. With his big, fat, lazy head, and smart little tail, he goes about the tank, apparently so happy, not minding confinement in the least, until he is gradually transformed into a frog, when he becomes a more serious member of the

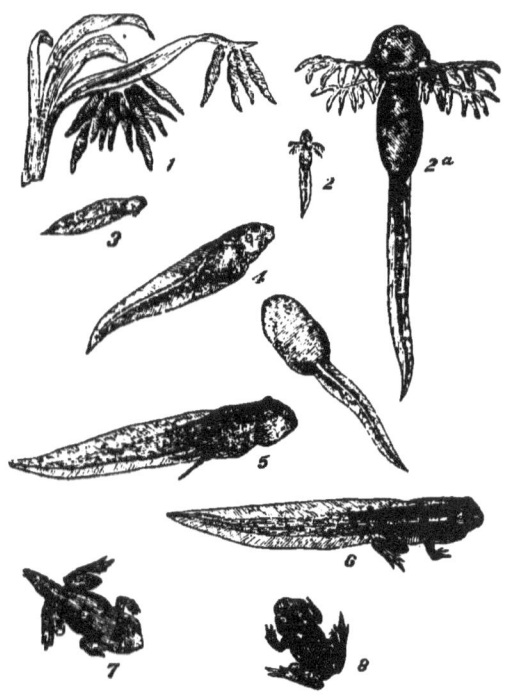

EIGHT STAGES OF THE DEVELOPMENT OF THE TADPOLE, from the recently hatched (1) to the adult form (8).

family, and might better be dismissed. But the change itself is very interesting to watch. In his transformations his last legs appear first, and his first or fore legs last. "But what about that lively tail?" Well, its wiggle-waggle has become gradually fainter, and moves in more sober mood than in his greener days; more surprising still, it appears to be van-

ishing!   Where it goes to it is hard to see, but finally it
disappears—actually all absorbed into the body, which is no
longer tadpole, but plain frog !   Poor "tad" used to be of

COMMON FROG (*Rana temporaria*).

some use, too, for he was a splendid scavenger, and helped
to clean the glass with his thick, fleshy lips.   Give the little
"tads" a home.

The caddis-worm, the larva of the caddis-fly, of the family of *Phryganida,* you will find on the bottom of running
brooks, nicely incased in his house of sticks, stones, and
sand, and sometimes of leaves.   He is an interesting little
creature, and sometimes builds his house in such a queer
way that you are sure to want to be better acquainted with
him.   A large number of their eggs—probably hundreds
of them—were once hatched out in one of my glasses ; and,
before the young were twelve hours old, each had built a
house over its tiny body, and was carrying it about almost
as well as the older members of the family.   They stick
these particles together by a cement which they secrete from
their own bodies, and the inside of this little tube or case is
covered with a silky lining, from which they stick out their

heads and fore-feet when moving about. Should the case when completed prove too light, so as to be inclined to float, they fasten on an extra stone to weigh it down.

Should you ask me which to choose, a salt-water or a fresh-water aquarium, I should reply, *Both*. "But this is not answering the question." Certainly not, nor can I answer it conclusively; for each has its peculiar advantages and special attractions. However, in point of rare and curious objects, undoubtedly the salt-water is the more prolific; but it is perhaps a little more difficult to manage, and some of the most desirable curiosities cannot be either so cheaply or easily procured as stock for the fresh-water tank. Probably an amateur would succeed best with the latter, at least until he has fully mastered both the principle and practice of the aquarian art. To stock a salt-water aquarium *handsomely* involves considerable time if undertaken independently, and expense if procured by purchase. There are, however, many very interesting objects which may be easily and cheaply obtained ; and of these I will first speak.

The water would better be *pure sea-water*, not taken in a brackish state at half-tide, when there is certain to be a large amount of foreign matter held in solution, and the saline property is weakened by the mingling of fresh-water streams. But if circumstances permit your going farther out, and you get your salt-water from the Bay or the East River, by all means consult the time-table, and let your dip for the precious fluid be as near full flood-tide as possible, or a little *before*, not after, for the former is when the sea-water is running in. I have many times used this water with perfect success ; and I have understood that specimens of water taken from around Hell Gate have been carefully analyzed and found to be *ninety-six per cent. pure,* and in density about the same as water taken from the locality of the light-ship at Sandy Hook.

There are formulæ for the manufacture of artificial sea-water which are well spoken of in the English works on this subject, but I have never tried any of them. I imagine there are conditions and occult combinations in ocean-water which no art can *fully* reach. I know, however, a resident of Cincinnati, Ohio, who has used, and is still using, artificial sea-water with very good results. He has a number of salt-water algæ and animals in a perfectly healthy condition in his tanks.[1] The principle upon which the marine aquarium is based and managed is the same as for the fresh-water, here-tofore explained.

The aëration is a very simple matter, for the means are abundant. The common sea-lettuce (*Ulva latissima*), which is washed up on every beach, is the best and really the only necessary vegetation for the salt-water aquarium. Nothing can be more beautiful either in color, texture, or form; and it will grow floating as well as rooted. A handful of the loose leaves picked up on the shore will, if clean, begin the work of aëration as soon as they are placed in the tank. The only objection to these loose fragments is, that they do not remain stationary. But we are not obliged to use them, for there are thousands of oyster-shells, pebbles, and small

---

[1] For the benefit of those who may live far away from the sea-shore, I will give Mr. Gosse's formula for making artificial sea-water, which is probably as good as any in use:

	Parts.
Common table-salt	81
Epsom salts	7
Chloride of magnesium	10
Chloride of potassium	2
Total	100

One pound of this mixture carefully dissolved in water, and then filtered, will make about three gallons of sea-water.

pieces of rock, to be found on any beach, covered with the younger forms of the same sea-weed, which can be arranged picturesquely and permanently in the tank, and which will supply oxygen freely. Other algæ may be added to beautify and vary the effect. Indeed, a perfectly gorgeous appearance may be produced by the many-colored and delicate algæ. It was formerly believed that the red, being mostly deep-sea plants, could not with safety be introduced into the aquarium ; but *experience*, that faithful teacher, has proved the unsoundness of that belief. As with the fresh, so with the salt-water tank, it is best to introduce your vegetation a few days before you stock it with animals. I will mention some of the many varieties of algæ that have grown in my aquaria at different periods, many of which, however, I do not recommend, as the aquarium is safer without them.

Besides the *Ulva latissima* already mentioned, we have the *U. linza, U. intestinalis, Enteromorpha compressa, Cladophora reflecta*, and the *Grenella Americana*, bright red, ribbon-like, and, if collected while it is young, very good for the tank ; *Porphyra lacinata*, royal purple, rich and beautiful, and as dangerous as it is pretty ; *Chondrus crispas*, commonly called Irish moss, the plant so extensively used in commerce, which is a valuable addition if young and small specimens are used ; *Rhodymenia palmata* and *R. laciniata*, both red plants. The *Solieria chordalis* is of a brilliant crimson color, with round, wire-shaped branches, growing profusely on shells and stones in a few fathoms of water ; this is one of the few that should never be dispensed with. *Dasya elegans* is one of the most ornamental of ocean-plants, of a soft, feather-like shape and texture, but not safe for the aquarium. *Ceramium fastigiatum* is one of the most beautiful objects to look at with a strong lens that can be imagined ; dried and pressed on paper, it is one of the choicest. *Callithamnium byssoidium* is not very hardy. *Polysiphonia parasitica* and *P. nigrescens* can be almost always found near the shore, and are about the first plants which you may expect to appear in your tank, whether you put them there or not; for some of the seeds or spores are nearly sure to be contained in the sea-water. They belong to a family of sea-weeds numbering over three hundred known varieties. *Cladophora arcta* is a fine, soft, silky green plant, growing in little tufts, appearing on the shore at early spring, and always tempts the amateur, saying by its seductive charms, " Try me;" but my advice is, *Beware*. *Ptilota elegans*, one of the pretty red weeds with beautiful feathered fronds, makes a fine contrast with the greens. *Delesseria sanguinea* and *D. Americana* are two of the most beautiful sea-weeds to be found upon our shore, and will do very nicely for the aquarium during the cold season.

Of specimens obtained in the Bermudas, I have had growing *Acetabularia crenulata, Caulerpa plumaris, Zonaria, Halimeda, Padina pavonia, Bostrychia scorpioides*, and *Penicillus capitates*, a very pretty tree-like form.

Supposing that we have our plants arranged and in good working (i. e., oxygen-producing) order, we are now ready for the animals, and will begin with the

MOLLUSKS AND CRUSTACEA.

The choice of mollusks is greater for the salt-water than for the fresh-water aquarium. Their use is the same. The common salt-water snails—as we shall have to call them— are to be counted by millions on almost every beach; they are good workers, and in my eyes beautiful creatures. They occupy themselves almost entirely with devouring the confervoid formations on the sides of the tank; but I know by observation that they are also useful scavengers of animal matter. If your fishes chance to have been over fed, the snails will eat up the fragments. The periwinkles (*Littorina palliata* and *L. rudis*), whose natural food is decayed vegetation, can be used with advantage. But the whelks (*Buc-*

cinum) are probably the most active of all, as they will consume all decayed animal matter as well as the confervoid growth. It is an interesting sight, if you have a proper glass, to watch these little creatures at work. With their scythe-like tongues they mow away the fine growth of algæ in perfect swaths, until the whole glass is swept clean; the shell of the whelk is a favorite dwelling for the smaller hermit-crabs, as described on page 100. The *Tritia trivittata* is a pretty spiral-shaped shell, found upon the sandy shores of Long Island; it is quite useful for the tank, and also ornamental. The *Anachis similis* and *Cerithiopsis terebralis* are almost invaluable for a small tank. Beware of the *Urosalpinx cinerea*, commonly called "the drill" (described on page 44); it is interesting, but may prove an expensive luxury to harbor. The scallop exceeds all in beauty, but its life is somewhat precarious in an aquarium; its richly-tinted, corrugated shell alone is an ornament, but the inmate is beautiful beyond description. The *Mytilus edulis* or common mussel, *Modiola plicatula* and *M. modiolus, Mya truncata, Venus mercenaria, Crepidula fornicata, Anomia glabra,* and *Argina pexata,* have already been described, and many more might be given; but the above-named number of mollusks is quite sufficient, and they are all available for the tank.

FIDDLER-CRAB (*Gelasimus vocans*).

Among *Crustacea*, I would mention the hermit-crab (*see*

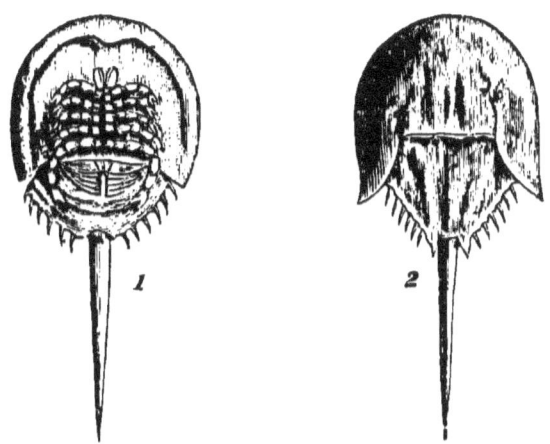

KING OR HORSESHOE CRAB.—1. Lower surface ; 2. Upper surface.

Chapter VIII.), which is most interesting. Any number of

HORSESHOE-CRAB IN TROUBLE (*Limulus polyphemus*).

specimens can be obtained among the rocks and sea-weed at

low tide. If well, the hermit-crab is full of life and activity, he runs along the bottom of the tank, and climbs the rocks, if there are any, with surprising rapidity. The whole race, however, belongs to the "fighting brigade." Just watch him! He is also a good scavenger. Do not use the fiddler-crab, for he is not wholly an aquatic animal; if you shut him up in your tank, he will revenge himself on you by dying and thus causing trouble. The common edible or blue crab, *if very small*, may be introduced "with care." The small lady or land crab, and the modest rock-crab, are acquisitions, being good-natured and harmless. And last, but not least, procure a few of those interesting little horseshoe-crabs (*Polyphemus Americanus*); they may be found half covered with sand on Coney Island.

The common sand or silver shrimp lives well in the aquarium, and is one of the most graceful creatures imaginable.

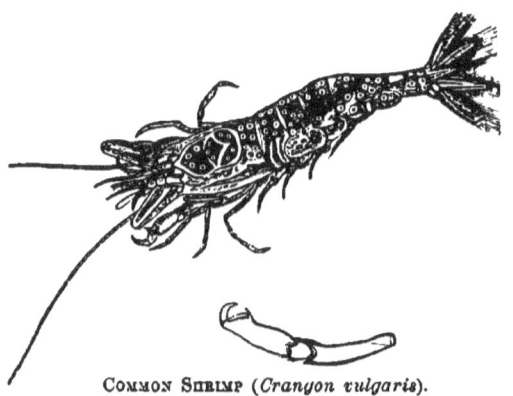

COMMON SHRIMP (*Crangon vulgaris*).

Remember that all crustacea *need feeding* regularly; a small piece of raw meat, cut fine, suits them as well as anything.

ZOÖPHYTES.

Here we enter into the exclusive domain of the salt-water aquaria, and upon the class of objects most difficult to obtain. First of this class are the *Actiniæ* or sea-anemones, described in Chapter II. They are properly called flower-shaped animals, though it was long doubted whether they *were* animals. There is a certain class of sea-anemones which can be obtained among the rocks when the tide is out, but only a practised eye would detect them; for, though they will be in full bloom when covered by the water, when deserted by the friendly wave they look only like slimy accumulations upon the rock. But when you have once learned to recognize them, there is no further difficulty—except to get them. This is not always a very easy matter, and practice alone will make you expert in detaching them from the rock. Indeed, this is always more or less hazardous, because it is at their base that they are most easily injured, and often fatally by the inexperienced. The best way when collecting them is to supply yourself with a hammer such as geologists use, and split off the piece of rock to which they adhere. A very pretty, small, white variety can be found at Gowanus Bay, and a deep, rich orange-colored one at Hell Gate. I have found them here so thick upon the rocks that the surface would be entirely covered. Sometimes they fix themselves upon a piece of brick, as many as two hundred young ones perhaps together; or the sole of an old shoe may be their resting-place. These will live well and multiply in the aquarium. The larger and finer anemones, and those of the most beautiful and varied colors, are obtained only in deep water, and generally by dredging. Some of these are truly gorgeous; but about the best suited for the tank is the *Actinia mesembryanthemum*, which comes from the Bermuda Islands, and can almost always be had at the aquarial store of Mr. Greenwood. Its color varies from a deep blood-red to a delicate scarlet. It never grows to more than an inch and a half in diameter, and multiplies readily in the tank during the winter. Sea-anemones can be safely carried in a basket, if packed in wet sea-weed.

Like all inmates of the aquarium, they should be regularly
fed, as no tank would naturally supply sufficient food such
as they require. I generally feed mine with pieces of clams,
but any finely-cut raw meat will do. If I wish to "treat
them" and give them a sort of *Thanksgiving dinner*, I get
some of those little crabs which are found in oysters; and
these they cannot resist even when not hungry. They remain closed for some time after feeding, to bloom, however,
after the food is digested with fresh vigor and beauty. I
think enough has been told about sea-anemones; like all
natural objects, only study and observation will enable you
thoroughly to appreciate them.

SERPULÆ AND HYDROIDS.

There is a class of remarkable animals known as annelids,
which are great acquisitions to the salt-water aquarium, but
which should be introduced with caution and watched with
vigilance. The proper care of them requires considerable
experience. The *Serpulæ* form contorted cellular tubes *en
masse*, to which fishermen give the name of "clinkers," supposing them to be pieces of clinkered coal which have been
acted upon by salt-water. These cells, however, are inhabited by gayly-colored, lively animals, as described fully in
Chapter IV. Another interesting and valuable specimen
for our tank is the *Tubularia indivisa*, unquestionably
one of the most beautiful things which the sea produces.
The groups formed by it can sometimes be found at Hell
Gate, but are not always conveniently situated so as to be
got at and removed with safety. To those only who explore
the bed of the ocean are all its wonders revealed ; but enough
of this class of objects has been described in these pages to
enable any one to make a fine selection for a sea-water aquarium.

THE MOST DESIRABLE FISHES.

First, as with fresh water, so with salt—the smaller the fish, the better.   Just brush your hand-net along the margin of that eel-grass, about a foot below the surface, and examine its contents without taking it wholly out of the water.   I suspected as much!   Those delicate, semi-transparent, silvery-striped *spearings* are indeed a treasure for your aquarium.   I fear, however, you will find them, as most exquisitely beautiful things are, but short-lived.   Expose them to the air as little as possible in transferring them to your " collection-pail," for they are very sensitive.   The common belief that they will die if they leave their natural element and breathe the air but for a moment, is, however, erroneous.   But " handle them tenderly, lift them with care," for they are worth it.

What else have you?   Ah! those chubby little fellows, with their bars of emerald on a shield of gold, are sheepshead (*Lebias*)—the best fish for the marine aquarium that I know of.   They are vegetarians, with an insatiable appetite, and forage freely on your sea-lettuce and the delicate algæ; but never mind, the *Lebias* is just as much at home in the tank as in the vasty deep, and you may keep him till he dies of old age; and he is beautiful as he is hardy.

And a young flounder, too!   Good again.   But if you mean to preserve him in health, you must indulge him with a sand-bank in which to lie and to burrow.   And an infantile blackfish!   Yes, he will do admirably; for he is a tough little chap, and is marked so prettily with black and white. A young bergall is a fine addition to your catch; he will glide in and out among your rock-work as independently and gracefully as if he owned the whole aquarium.   And there's a pipe-fish!   Well, he's curious if not handsome : drop him in the pail.   But that frost-fish—let him go.   If you transfer him to the handsomest aquarium that ever was

made, he will not content himself, but will sulk and refuse
to eat, and will consequently die, as you, dear reader, would
if persisting in such a refusal.

AMERICAN FLOUNDER (*Platessa plana*).

My pet of all pets I leave to the last—the little sea-
horse, described in Chapter IX. If you once have one
of these under your eye, you will never wish to be with-

SEA-HORSE LOOKING FOR FOOD.

out them; and yet they are delicate, and there is no telling how long—or rather how short—a time you may be able to keep them. But "a thing of beauty is a joy forever," to memory if not to sight; so, if you *can* obtain one, don't deny yourself the pleasure for fear it will not last.

> " Under bright rainbows of perpetual showers,
>   God's gardens of the deep
>   His patient angels keep;
> Gladdening the dim strange solitude
>   With fairest forms, and thus
>   *Forever teaching us.*"

And now it is time to close these pages, and to say to the reader who has accompanied me to the end: Farewell. Or, rather, shall it be *au revoir ?* May I not hope that some who have perused this little book will have become sufficiently interested in the objects described, not only to wish to possess them, but to obtain them personally — seeking such as may be found in our vicinity in their native retreats? Shall we meet then, reader, perchance on the Coney Island shore (*now* Manhattan Beach), or on some rocky projection of the East River, or, mayhap, on the Hackensack or Passaic, and, in the spirit of brotherhood which the study of Nature engenders, compare our respective "finds," and rejoice with each other over some new treasure discovered? If not, I shall hope that at least my experience and suggestions may be of use to those interested in aquarial objects; that they may encourage some, who have failed, to "try again;" and that it may save from loss and disappointment some amateur just beginning to make collections.

To the more advanced aquarianist, possibly my *personal experience* may not prove wholly useless; and if the reader, of any class, has enjoyed the perusal of this book, and learned to love my little pets, I ask no further recompense. 11

APPENDIX.

HOW TO COLLECT AND PRESERVE ALGÆ.

THE best manner to gather marine plants to be sent abroad is to take them from the rocks at low water, and dry them in the *shade*. Then pack them in paper or muslin bags, and mark the locality, day of the month and year, in plain letters. In gathering plants, always take them with the roots when you can; and never, on any account, let the plants (if marine) be placed in *fresh water*, in which they lose their color and soon decay. It is a good plan to gather *in mass* the plants as they are thrown up from the sea and cast ashore, and examine them carefully at leisure, as many rare and beautiful specimens grow in deep water. If you wish to preserve them on the spot, provide yourself with paper and small pieces of white muslin, and float them out in sea-water, taking care to cover each specimen with a piece of muslin, and over this a sheet of blotting-paper before putting on any pressure. In this manner you can preserve many specimens together without injury. It is well to prepare the delicate specimens on the spot, as they decay very soon. The large, coarse specimens are always best preserved by being hung up to dry in the shade, and when nearly dry they will be ready to pack away in paper or muslin bags. Do not discard specimens that are coarse and ugly, as they often have minute parasites attached to them of great interest to the botanist; and frequently plants that look like decayed specimens are rare and valuable to the algologist. Collect everything in the shape of a plant that grows in the water, and preserve them, as they will be of interest.

INDEX.

THE END.

THE WORLD OF WATERS.

The Commercial Products of the Sea;

Or, Marine Contributions to Food, Industry, and Art. By P. L. SIMMONDS. With Thirty-two Illustrations. 1 vol., 12mo, cloth. 484 pages. Price, $1.75.

"It is certain that a large part of our race would have to perish but for the countless forms of life which swarm in the depths of the sea. Mr. Simmonds has brought together the statistics as to the extent of the various large fisheries of the world, whether they supply food or objects of industrial value. Of the most exquisite and highly-valued product of the sea, the pearl, and of the methods of obtaining it, and managing the pearl-oyster beds, Mr. Simmonds gives a careful account. To the sponge, coral, amber, turtle, and other fisheries, he devotes the same detailed attention."—*N. Y. Sun.*

"Filled with important and interesting facts concerning marine contributions to food, industry, and art."—*Boston Journal.*

"A vast amount of information is here collected, and the general reader, as well as the specialist, will find the work full of interest."—*Buffalo Daily Courier.*

"The Multitudinous Seas."

With Illustrations. By S. G. W. BENJAMIN. Forming Number 23 of Appletons' "New Handy-Volume Series." 18mo. Paper, price, 25 cents.

"Another view of the inexhaustible fascinations of the deep is displayed in 'The Multitudinous Seas,' by S. G. W. Benjamin. Free from statistics, and written in a fluent and picturesque style, by one who has traveled far and near over the face of the waters, it deals little with subjects of industrial importance, but enables one to gain, in an easy and attractive way, a new insight into the myriad strange aspects of the ocean. St. Elmo's lights, water-spouts, submarine volcanoes, icebergs, cyclones, the beautiful and curious creatures found floating on the surface of the sea, are only a part of the themes on which it discourses."—*N. Y. Sun.*

Figuier's Ocean World:

A Descriptive History of the Sea and its Inhabitants. With 425 beautiful Illustrations. Carefully revised. By E. PERCIVAL WRIGHT, M. D. 1 vol., 12mo. Cloth, $3.00; half calf, $5.00; full calf, $6.00.

D. APPLETON & CO., PUBLISHERS, 549 & 551 BROADWAY, N. Y.

HEALTH PRIMERS.

EDITED BY

J. LANGDON DOWN, M. D., F. R. C. P.
HENRY POWER, M. B., F. R. C. S.
J. MORTIMER-GRANVILLE, M. D.
JOHN TWEEDY, F. R. C. S.

THOUGH it is of the greatest importance that books upon health should be in the highest degree trustworthy, it is notorious that most of the cheap and popular kind are mere crude compilations of incompetent persons, and are often misleading and injurious. Impressed by these considerations, several eminent medical and scientific men of London have combined to prepare a series of HEALTH PRIMERS of a character that shall be entitled to the fullest confidence. They are to be brief, simple, and elementary in statement, filled with substantial and useful information suitable for the guidance of grown-up people. Each primer will be written by a gentleman specially competent to treat his subject, while the critical supervision of the books is in the hands of a committee who will act as editors.

As these little books are produced by English authors, they are naturally based very much upon English experience, but it matters little whence illustrations upon such subjects are drawn, because the essential conditions of avoiding disease and preserving health are to a great degree everywhere the same.

VOLUMES OF THE SERIES.

In square 16mo volumes, cloth, price, 40 cents each.

For sale by all booksellers. Any volume mailed, post-paid, to any address in the United States, on receipt of price.

D. APPLETON & CO., PUBLISHERS,
549 & 551 BROADWAY, NEW YORK.

SOCIAL ETIQUETTE OF NEW YORK.

CONTENTS: The Value of Etiquette — Introductions — Solicitations — Strangers in Town—Débuts in Society—Visiting, and Visiting-Cards for Ladies—Card and Visiting-Customs for Gentlemen—Morning Receptions and Kettle-Drums—Giving and attending Parties, Balls, and Germans—Dinner-giving and Dining out—Breakfasts, Luncheons, and Suppers—Opera and Theatre Parties, Private Theatricals, and Musicales—Etiquette of Weddings—Christenings and Birthdays —Marriage Anniversaries—New-Year's-Day in New York—Funeral Customs and Seasons of Mourning.

18mo. Cloth, gilt edges, price, $1.00.

"This little volume contains numerous hints and suggestions, which are specially serviceable to strangers, and which even people to the manner born will find interesting and useful. Perhaps the best part of it is in what it does not say, the indefinable suggestion of good-breeding and refinement which its well-written pages make."—*New York Evening Express.*

"A sensible and brief treatise, which young persons may profitably read."— *New York Evening Post.*

"Everything which refines the habits of a people ennobles it, and hence the importance of furnishing to the public all possible aids to superior manners. This book will undoubtedly meet the needs of a large class."—*Boston Evening Transcript.*

"A frank and sensible epitome of the customs of good society in the first city of America. It admits the existence and need of certain rules of social behavior, and then in a kindly and decorous manner points out how to conform to the best usage."—*Boston Commonwealth.*

"A very sensible and—if we may say it of a book—well-bred volume. It gives the rules that are observed in the metropolis. These sometimes seem artificial, but they are usually founded on reason."—*Hartford Courant.*

"This is a timely work. For years our people have followed the habits of the older nations. In this young republic it can not be expected that the same rules exist as we find abroad. This work is very complete, and is easily carried in the pocket to read at odd intervals."—*Albany Sunday Press.*

"The statements are exact and simple, and cover all that any reader is likely to desire. The work will convey positively useful and reliable instruction that can not always be reached otherwise."—*Philadelphia North American.*

D. APPLETON & CO., PUBLISHERS, 549 & 551 BROADWAY, N. Y.

RAMBLES IN WONDERLAND;

OR,

*Up the Yellowstone, and among the Geysers and other Curiosities of the
National Park.*

- By EDWIN J. STANLEY.

WITH MAP AND TWELVE ILLUSTRATIONS.

Large 12mo. Paper cover, price, 75 cents; cloth, $1.25.

"The natural wonders of the Great West, and especially those of the
Yellowstone region, have been frequently described. But it can be safely
said that, however familiar they may have become, either through books
or by travel there, every one will find these sketches of them well worth
reading. It is a most impressive volume; and this comes from the fact
that the author gives a plain and clear description, and does not attempt
to portray the wonder or the admiration which he himself felt. The result is, that the grandeur of the objects themselves reaches, directly and
naturally, the soul of every reader. We commend the volume as one
which, in the first place, has an abundance of things which every American, at least, ought to know, and one which, in the second place, is un-
usually readable."—*N. Y. Churchman.*

"Mr. Edwin J. Stanley has made a book with the title 'Rambles in
Wonderland' out of his notes and letters written during a season of
travel up the Yellowstone River and through the Yellowstone Park. The
book pretends to no special literary excellence, but is briskly written,
and may be read with interest. Some of its descriptions are very graphic
and picturesque, and, with its excellent illustrations, it is a travel-sketch
of much interest and value."—*N. Y. Evening Post.*

"The famous cañons, the hot springs or geysers, the National Park,
the Indian agencies, the tribes of the Sioux, Crows, and other aboriginals;
Indian fighting, the massacre of pleasure-parties in the National Park,
hunting, fishing, and the usual adventures of travel in a wild country, are
among the subjects treated."—*N. Y. Home Journal.*

"An account of the summer rambles of a Methodist preacher in the
wondrous Yellowstone region. The numerous chapters are vivid pictures
of the journey to and through that enchanted land."—*N. Y. Christian
Advocate.*

"This is a well-printed book of 179 pages, by a worthy and useful
Southern Methodist preacher—one of our brethren on the far frontier of
the new Northwest. There is much in the book to interest and instruct.
It is pleasant reading for a man; it would delight a boy with any soul in
him."—*Macon* (Ga.) *Wesleyan Christian Advocate.*

D. APPLETON & CO., 549 & 551 Broadway, New York.